UK price
£8.99

COMBAT AIRCRAFT

F-15

a Salamander book

Published by Salamander Books Limited
LONDON

A Salamander Book

Published by Salamander Books Ltd,
129-137 York Way,
London N7 9LG,
United Kingdom.

This edition © Salamander Books Ltd, 1992

ISBN 0 86101 677 7

Distributed in the United Kingdom by
Hodder & Stoughton Services,
P.O. Box 6, Mill Road,
Dunton Green, Sevenoaks,
Kent TN13 2XX.

Credits

Project Manager: Ray Bonds

Editors: Bernard Fitzsimons, Tony Hall

Designed by: Tony Dominy, Rod Teasdale.
This 1992 edition adapted by Studio Gossett.

Diagrams: Arka Graphics © Salamander
Books Ltd. .

**Three-view, cutaway drawing and colour
profiles:** © Pilot Press Ltd.

Jacket: Brian Knight

Filmset by Tradespools Ltd, SX Composing
Ltd.

Colour reproduction by Bantam Litho Ltd,
Scantrans PTE Ltd, Singapore

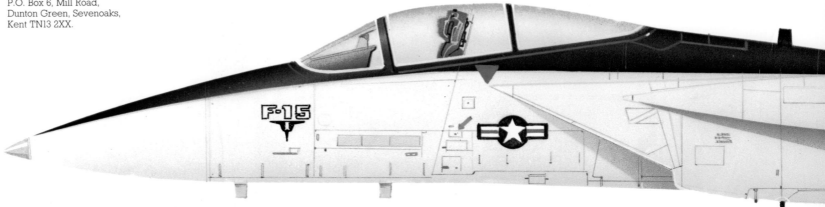

Contents

Acknowledgements

The author and editors are grateful to
the following individuals. Geoffrey
Norris, Karen Stubberfield and Jeffrey
Fister of the McDonnell Douglas
Corporation; Major Dito Ladd and
Colonel Wendell H. Shawler. For the
1992 edition: Col. Arthur Forster, Maj.
Garian Perugini, Maj. Tom Griffith,
Capts Don Watrous, Dave Harkins, Eric
Schnaible, Paul Parkham, 2 Lt. Karen A.
Finn, S. Sgt Steve Smith, Sgt Henry
Clark, Dave Peterson, Jay Barber, Dave
Wilton, David Oliver, Lee Whitney,
Kearney Bothwell, Chuck Suits, Jim
Swison and Gill Vale.

The Powerplant chapter is based on
the equivalent chapter in the
companion volume *F-16 Fighting
Falcon*. The author is further grateful to
Don Parry and Bernard Fitzsimons who
contributed the chapters on Avionics,
and Structure and Performance and
Handling.

Authors

Michael J. Gething is Managing Editor
of the Whitton Press *Defence* series of
publications and Editor of *Defence*.
Before joining the staff of the magazine
in 1976 he had worked for the Defence
Operational Analysis Establishment at
West Byfleet, and had been Assistant
Editor of the Royal Aeronautical Society
magazine, *Aerospace*.

Fascinated by aviation history since
his youth, he began his career as an
aviation journalist by writing on scale
modelling subjects. Although he has
less time to build models, he keeps in
touch with the scene through the
International Plastic Modellers Society.

In his spare time he is a squadron
officer with the Bracknell Squadron of
the Air Training Corps, having been
commissioned in the Royal Air Force
Volunteer Reserve (Training Branch) in
1972. Among his other books are *NATO
Air Power in the 1980s*, *Warsaw Pact Air
Power in the 1980s* and *Military
Helicopters*.

Paul Crickmore
Author of the 1992 updated edition,
Paul Crickmore began as an aviation
journalist and photographer after 13
years as an air traffic controller. He has
previously written books on the SR-71,
RF-4C and U2/TR-1 as well as
numerous articles for the aviation
press, including such publications as
the *RAF Yearbook*, *Air Forces Monthly*
and *Air International*.

Introduction

Fighter aircraft have always held a special fascination, from the apparently flimsy scout biplanes of the First World War, through the legendary Spitfire, Mustang and Zero of the Second World War and the F-86 Sabre that fought over Korea to the F-4 Phantom of the 1960s. Even compared with such predecessors the F-15 Eagle sets a completely new standard, and one that only today is even beginning to be matched by the MiG 29 Fulcrum and Su 27 Flanker.

The Eagle's exceptional capabilities are the product of a combination of features. Two powerful F100 engines provide the 50,000lb (22,700kg) of thrust for record-breaking climb and acceleration; an airframe refined through thousands of hours of computer studies and wind-tunnel tests translates raw power into outstanding manoeuvrability; digital avionics and a powerful long-range radar enable targets to be detected at ranges of up to 100 miles (160km); and new versions of the medium-range radar-guided Sparrow and short-range heat-seeking Sidewinder, along with a high-speed Gatling gun, form the Eagle's deadly claws. All these aspects are dealt with in authoritative detail in the following pages, along with accounts of the type's performance and handling qualities, its deployment and combat record with the US Air Force in the Gulf and the Israeli Defence Force/Air Force over Lebanon, and the development of the specialized F-15E Strike Eagle for ground-attack missions.

The one major drawback with this exceptional aircraft is its price: only Israel, Japan and Saudi Arabia have so far found the wherewithal – and the official approval – to acquire their own F-15s, and not even the US Air Force can afford all the Eagles it would like. However, it was inevitable that the best fighter in the world, would also be one of the most expensive, and in terms both of capability and of potential for development, it would seem that the bill is equalled by the stature of the product.

Development

Confronted with a new generation of Soviet combat aircraft in 1967, while it was fighting in Vietnam with the 'second-hand' F-4 Phantom, the US Air Force began serious work on a new fighter of its own. Designated FX, and intended to provide air superiority in friendly and hostile airspace alike, the new design was to be optimized for combat, with the power and agility to overcome any current or projected Soviet opponent. In the resulting F-15 Eagle, with its unequalled combination of performance, firepower and sophisticated avionics, the USAF believes it has such a machine.

There are two adages which apply to the McDonnell Douglas F-15 Eagle: 'if it looks good, it is good' and 'if you design a good fighter, it can be adapted to other roles successfully'. The basic objective of the F-15 programme was, according to Major General Benjamin N. Bellis, F-15 System Program Director, "to efficiently acquire a fighter capable of gaining and maintaining air superiority through air-to-air combat". Although the Eagle has yet to be flown in action by the US Air Force it has fulfilled its purpose, again according to Maj Gen Bellis of being "a high performance, extremely agile aircraft to meet the projected threat of the late 1970s and early 1980s" at the hands of Israeli pilots over the Lebanon. The air-to-ground capability built into the original design is now being enhanced to produce the F-15E 'Strike Eagle', and the type is also being procured as an air defence interceptor to replace the F-106 Delta Dart.

The F-15 Eagle is the first air superiority fighter to stem from USAF requirements since the F-86 Sabre of 1948. The previous TAC (Tactical Air

Below: Silhouetted against the setting sun, a pair of F-15As in the skies they were designed to dominate.

Command) fighter was the F-4 Phantom II, designed for the US Navy and, together with the A-7D Corsair II, forced on the USAF by the circumstances of the Vietnam War. Having to procure Navy aircraft was anathema to the Air Force, and despite the success of both Phantom and Corsair, when the time came for a new tactical fighter the Air Force was determined that it should be of their own choosing.

New Soviet fighters

The genesis of the Eagle can, perhaps, be traced back to a Russian airfield at Domodedovo, near Moscow in July 1967. At an airshow there, in front of the world's press, the Soviet Union unveiled a new generation of combat aircraft. Of particular note were a swing-wing fighter, condenamed Flogger by NATO, and a high-speed, twin-fin fighter codenamed Foxbat, both from the famous Mikoyan-Gurevich (MiG) design bureau. Later information was to identify the Flogger as the MiG-23 and the Foxbat as the MiG-25, while a later ground-attack version of the MiG-23 was designated MiG-27 Flogger D.

The MiG-23 Flogger was a single-seat air-combat fighter powered by a single Tumansky R-29 afterburning tur-

bojet with a Mach 2.2 capability. Armed with a twin-barrel 23mm GSh-23 cannon and four air-to-air missiles – later known to comprise a pair each of AA-7 Apex and AA-8 Aphid AAMs – its range of 1,200 miles (1,930km) and service ceiling of 61,000ft (18,600m) made it a potent fighter. It was to supplement the MiG-21 Fishbed series of fighters, then in their second and now in the third stage of their development.

The MiG-25 Foxbat was also a single-seater, but powered by a pair of Tumansky R-31 afterburning turbojets. Initially a missiles-only interceptor, with a pair of AA-6 Acrid (one infra-red homing and one radar homing) plus a single AA-7 Apex and a single AA-8 Aphid, it carried a high-power radar codenamed Fire Fox. It has a Mach 2.8 capability, could fly 1,610 miles (2,580km) and had a service ceiling of 80,000ft (24,400m). Initially the design was thought to have been a counter to the American B-70 Valkyrie bomber cancelled in 1961, but the Foxbat was continued as an air superiority and reconnaissance fighter. The implication was obvious to the Pentagon: here was a fighter that could prove immune to the standard USAF fighter of the day, the F-4 Phantom.

Above: Air superiority blue and dayglo orange paint scheme on one of the pre-production F-15s during the flight test programme.

The FX study

Work on a new air superiority fighter had begun within the USAF as a general feeling of need for an aircraft in the best traditions of the P-51 Mustang and F-86 Sabre. This was in the early 1960s, and by April 1965 the USAF fighter lobby were looking at a Fighter Experimental (FX) type. In October 1965 the USAF asked for funding of full scale studies, and two months later issued a Request for Proposals (RFP) for a Tactical Support Aircraft. The Concept Formulation Study (CFS) which came out of the RFP went to Boeing, Lockheed and North American Rockwell in March 1966, the McDonnell Aircraft Company (MCAIR) being one of the losers at this stage. However, none of the submitted designs was considered further, mainly due to the aerodynamic configurations and by-pass ratio of the powerplants.

From mid-1966 to autumn 1967 activity on the FX was minimal, although the USAF maintained its own CFS team in being until autumn 1968. The impact of the Domodedovo revelations was felt,

and in August 1967 a second RFP for a CFS was issued. This time the words Tactical Support Aircraft were changed to one – Fighter – and this time MCAIR, along with General Dynamics (formerly Convair) were awarded the six-month study.

Among the objectives set by the USAF was a speed range of Mach 1.5 to 3.0. General Dynamics offered both a variable geometry and a fixed-wing FX: MCAIR recommended a fixed wing, twin engines and a single crewman. This second CFS was completed in May 1968, and in September of that year the FX Concept Development was authorized. In the same month the RFP for Contract Definition stage was offered to the aerospace industry, and MCAIR, along with Boeing, Fairchild Hiller, General Dynamics, Grumman, Lockhead, Ling Temco Vought and North American, bid for the contract. By De-

cember 1968 only MCAIR, Fairchild Hiller and North American were in the running.

By now the FX was designated F-15 and the three contenders were hard at work. A Development Concept Paper issued by the USAF defined the overall parameters of the design, and justified it against pressure from the US Navy to take a modified version of their VFAX/F-14 on four counts: it would be a single-seat, fixed-wing, twin-engined fighter of approximately 40,000lb (18,000kg);

Top left: The 60,000lb (27,000kg) variable-geometry proposal produced by MCAIR in early 1968.

Above left: slightly later design study with fixed delta wing.

Top: Variable-camber leading edges were envisaged for this wing form developed in early 1969.

Above: Wooden mock-up of the F-15 used for NASA wind-tunnel tests.

there would be no competitive fly-off, as this was not thought desirable; the VFAX was not considered a suitable replacement for the F-4E Phantom, nor could the F-4E be modified to meet the threat; and an air-to-ground capability

would be included, but only as an offshoot of the primary air-to-air role.

Initiated under the Total Procurement Package, the programme thus left a large volume of work for the contractors bidding. No hardware competition

Right: The first F-15A prototype airborne for the first time from Edwards AFB on July 27, 1972.

Below: A month earlier, the same aircraft was photographed before the roll-out ceremony at St Louis.

Above: The third development F-15A, used to test the avionics, is seen here in the markings of the USAF Flight Dynamics Laboratory Advanced Environmental Control System.

F-15 Project demonstration milestones

Preliminary design review	Sep 70
Radar contractor selection	Sep 70
Critical design review	Apr 71
Avionics review	June 71
Major sub-assembly tests	Jun 72
Engine inlet compatibility	Mar 72
First flight	Jul 72
Bench avionics complete	Sep 72
First aircraft performance demonstration	Sep 72
First airborne avionics performance	Dec 72
Fatigue test to reach one lifetime	Jan 73
Static test 2 critical concluded	Jan 73
Armament ground test	Jun 73
1g flight envelope	Aug 73
Fatigue test to reach 3 lifetimes	Dec 73
USAF evaluation summary	Dec 73
Equipment qualified	Mar 74
Category II aircraft and equipment in place	Mar 74
Training equipment in place	Oct 74
Fatigue test to reach 4 lifetimes	Oct 74
External stores flutter and release	Aug 74
AGE equipment in place	Oct 74
Category I flight tests complete	Nov 74
First aircraft delivered to TAC	Nov 74

meant an enormous amount of paper studies and documentation, while the contract itself had to include tooling, development, testing and production. Later, in response to criticism from Congress and the public over cost overruns on the C-5A Galaxy and F-111 programmes, the USAF worked in demonstration milestones which the contractor had to meet before receiving the next stage of funding. These are listed in the accompanying table.

The bids were made by June 1969, and from July to December the USAF's Aeronautical Systems Division made their evaluation. Selection of the McDonnell Douglas bid was announced on December 23, 1969. Major General Bellis announced in Washington the next day that in the technical, operational, management and logistic areas McDonnell Douglas had been placed first. In addition he went on to say that the MCAIR price had been lowest of the three. Selection of the winning contractor was the responsibility of Secretary of

the Air Force Robert G. Seamans Jr, after hearing presentations from the F-15 source selection evaluation board and the source selection advisory council, neither of which made specific recommendations. Secretary Seamans' choice was favourably received throughout the USAF.

The initial contract called for 20 development aircraft: a preliminary batch of 10 single-seat F-15A (71-0280–71-0289) and a pair of two-seat trainer TF-15A (71-0290 and 71-0291, later redesignated F-15B) Category I versions; and eight Category II full scale development (FSD) aircraft in single seat F-15A form (72-0113–72-0120). The FSD batch were closely matched to the production configuration.

The first F-15 was rolled out officially at MCAIR's plant at St Louis on June 26, 1972 with due ceremony. In July it was taken apart, loaded into a USAF C-5A Galaxy transport and airlifted to Edwards AFB, California. There it was reassembled, checked out and prepared

Above: The second development TF-15, later redesignated F-15B, was flown for the first time on October 18, 1973, and has subsequently been seen in a variety of configurations.

for its maiden flight. On a typical Californian clear day, with blazing sunshine, MCAIR's Chief Test Pilot Irving Burrows took the first Eagle (USAF serial 71-0280) into the air on July 27, 1972.

The Eagle's missions
In the terms of the FX Development Concept Paper (DCP), the F-15 is "optimized for counter-air missions" operating as part of TAC. These missions, which come under the general heading of air superiority, include escorting friendly strike forces over enemy airspace, making fighter sweeps ahead of such a strike force, combat air patrol between friendly strike aircraft and enemy bases, and tactical air defence of friendly territory.

According to the DCP, the most difficult of these roles is combat over enemy airspace, where "the counter-air fighter must protect the strike force from enemy fighters while under the disadvantage of being in the enemy's GCI network and exposed to potential attack from his fighters, SAMs and AAA". It is no surprise, therefore, to learn that the DCP calls for the F-15 to be "superior in air combat to any present or postulated Soviet fighters both in close-in, visual encounters and in stand-off or all-weather encounters".

According to the DCP assessment, neither an improved F-4E, with new wings and engines, or a version of the VFX (F-14 Tomcat) for which a contract was placed in February 1969, were considered able to meet the FX requirement. USAF politics aside, the configuration as a carrier aircraft for the basic role of fleet air defence and its consequent cost ruled out the VFX. However, an 'escape clause' was written into the DCP, which considered a

Left: An August 1975 view of the second F-15B, seen here carrying out trials with the Fuel and Sensor Tactical (FAST) pack conformal fuel tanks on the intake sides.

F-15 pre-production aircraft flight test roles

Serial	First flight	Function	Serial	First flight	Function
71-0280	Jul 27, 1972	Open flight envelope; explore handling qualities; external stores carriage	71-0285	May 23, 1973	Avionics tests; flight control evaluation; missile fire control
71-0281	Sep 26, 1972	F100 engine tests	71-0286	Jun 14, 1973	Armament and fuel stores tests
71-0282	Nov 4, 1972	Avionics development; calibrated air speed tests	71-0287	Aug 25, 1973	Spin recovery, high AOA and fuel system tests
71-0283	Jan 13, 1973	Structural test airframe	71-0288	Oct 20, 1973	Integrated aircraft/engine performance tests
71-0284	Mar 7, 1973	Internal gun, external fuel jettison and armament tests	71-0289	Jan 16, 1974	Tactical EW system, radar and avionics evaluation

Below: F-15B 71-0291 in the Bicentennial paint scheme worn by this much repainted Eagle during 1976 as part of the celebrations of 200 years of American independence.

further examination when "revised, detailed information on the tradeoffs between VFX-2 (with new engines and lighter avionics) and FX will be available to enable a decision as to whether to pursue one or both of these aircraft".

It is further interesting to note that the Eagle itself was later considered for adoption by the US Navy. This occurred in July 1971, when the Secretary of Defense asked the Navy to investigate the possibility of an F-15N, via the Systems Program Office. A minimum modification study by MCAIR to equip the Eagle for carrier operations increased the weight of the aircraft by some 2,300lb (1,043kg). The Navy Fighter Study Group III then did their own appreciation, disregarding the MCAIR data, which resulted in further weight increases and the addition of AIM-54 Phoenix AAM. In this configuration the weight and drag sent the performance down and the costs up to a level where the F-15N was deemed unacceptable.

The navalized Eagle made one more appearance before being finally rejected. Investigative testimony before the Senate Armed Services Committee's ad hoc Tactical Air Power subcommittee started new discussions on a modified Eagle for the Navy. This was in March 1973, at a time when the F-14 Tomcat programme was under pressure. Deputy Secretary of Defense Packard wanted to look at a lower-cost F-14, the F-15N and improved F-4s as alternatives for the Navy mission. The result was the formation of Navy Fighter Study IV, where these alternatives were discussed, and out of it came the concept of the Naval Air Combat Fighter to partner the F-14 in a maritime equivalent of the Air Force F-15/F-16 hi-lo mix. Ironically, MCAIR was selected as prime contractor on this programme, which involved navalizing the Northrop YF-17 as the F-18 Hornet.

Each of the ten pre-production Eagles was allocated a specific task in the flight test programme. Their roles were as outlined in the table. The first two-seater, 71-0290, was first flown on July 7, 1973, and was used for the two-seater evaluation. The second two-seater was flown on October 18, 1973. The first two-seater was slotted between the seventh and eighth aircraft, and since then every seventh aircraft built has been in two-seat configuration.

Right: The Bicentennial Eagle in flight, sporting the flags of the many countries where it had been demonstrated below the canopy.

Below: Back to a more familiar colour scheme, but still sporting an array of flags below the canopy, 71-0291 touches down with speed brake deployed.

Right: 72-0118, the sixth of eight full-scale development aircraft produced for the USAF's own Category II flight test and evaluation programme.

The first 12 Eagles were allocated to Category I of the test programme (now known as Contractor Development, Test and Evaluation) under the manufacturer's test pilots. The eight FSD aircraft allocated to Category II testing (now the Air Force Development, Test and Evaluation) were flown by a USAF Joint Test Force of Air Systems Command test pilots and Tactical Air Command fighter pilots. Category III of the test programme (now the Follow-on Operational Test and Evaluation) was conducted by the USAF. Later some of the Category I aircraft were passed on to the USAF.

By October 29, 1973, when the 1,000th test flight was flown by an F-15, 11 of the 12 Category I airframes were flying. Those aircraft had expanded the flight envelope of the Eagle to a speed of Mach 2.3 and a height of 60,000ft (18,300m). While development proceeded reasonably smoothly, it was not without its problems, and MCAIR Chief Test Pilot Irving Burrows and Colonel Wendell Shawler, the USAF's Director of the Joint Test Force 'went public' on a number of these in a joint lecture to the Society of Experimental Test Pilots in 1973.

The first problem discussed related to the stick force per g value. As a result of simulator experience prior to the first flight, it was thought that the aircraft might not be as nimble as expected, simply because of the stick forces. As designed, it was thought that these stick forces would be comfortable for manoeuvring, while not being so low as to suggest the chance of a pilot-induced oscillation or aggravate high g sensitivity. There are two sets of controls: a conventional hydromechanical system, and a Control Augmentation System (CAS). Both are capable of flying the aircraft, though the former was considered a back-up system in the event of the CAS failing.

The hydromechanical system determines the basic control deflections,

Below: An early production F-15A in the markings of the 58th Tactical Fighter Training Wing, based at Luke AFB, Arizona, and wearing the standard Compass Gray two-tone paint scheme.

Above: Underside view of the 43rd production F-15A, the 61st of all single-seat Eagles.

Below: One of two F-15Bs which, along a pair of F-15As, all from the 58th TFTW, were painted in a special attitude-deception three-tone grey camouflage scheme devised by aviation artist Keith Ferris.

Right: F-15Cs on the McDonnell Douglas production line in June 1981, when over 600 Eagles had been delivered to the USAF.

Right: F-15Cs on the McDonnell Douglas production line in June 1981, when over 600 Eagles had been delivered to the USAF.

while the CAS operates over the hydro-mechanical system and modifies the control surface deflections which provided aircraft response in line with the stick position. In the mechanical longitudinal system a spring cartridge provided the linear force gradient. There was much debate as to what was the optimum setting for the cartridge, but as the evidence was based on simulator experience only there was some reluctance to make changes. Besides, it was thought unwise to lighten the stick forces too much in case the aircraft was accidentally overstressed. Thus with a stick force (ie the pressure required to be applied to the control column in order to move the control surfaces) of 3.75lb (1.7kg) per g, manoeuvres in excess of 6.5g required some 25lb (11.3kg) of stick force to initiate a response to control demand. In such high-g situations this meant some considerable strain on the pilot.

Modified stick forces

Early flight testing confirmed the initial simulation findings. With the CAS off the manoeuvring forces were too heavy for a fighter with the inherent capability of the F-15. With the CAS on these forces were more comfortable, but there was room for improvement. A dual-gradient longitudinal spring cartridge was evaluated, and found to be an improvement, while modifications were also made to the CAS pitch computer, enabling a satisfactory match between the CAS and the hydromechanical system. Flight testing of these fixes confirmed their suitability, and were incorporated in production aircraft. Forces around the neutral position are considered 'comfortable' at all speeds within the flight envelope, and there was no excessive longitudinal sensitivity or trend towards oscillation with the CAS either on or off. Pilots can now fly 6g manoeuvres one-handed with no trouble.

The lateral sensitivity of the Eagle also came in for some criticism. Although, again, it was recognized on the simulator, it was partially masked by the lack of physiological cues. The original specifications called for rolling capabilities in excess of previous accelerations. These could only be achieved by using a large amount of lateral control – quickly. This meant that lateral control surface deflections were 'sudden and big'. While the ailerons, with plus or minus 20deg of travel, are mechanically served, the differential stabilator (all-moving tailplane) is connected to both the mechanical and CAS circuits.

Normal smooth manoeuvring was highly responsive, but comfortable. However, any sudden small lateral movements of the stick, such as might be expected during formation flying, gun-tracking or air-to-air refuelling, caused an undesirable jerkiness resulting in a possible pilot-induced oscillation. While not as simple to correct as was the longtitudinal system, a two-point solution was arrived at. A dual-gradient force system coupled with a higher setting on the CAS transducer prevented the CAS from augmenting roll commands at small stick deflections was the first fix; the second involved a modification to the CAS in order to

Right: Tanker's eye view of a two-seat Eagle, with the refuelling boom positioned in the F-15's receptacle. In-flight refuelling has enabled Eagles to fly non-stop the 7,000 miles (9,620km) between Okinawa and Florida.

Above: Three brand new F-15As in flight early in 1976 before delivery to the 1st TFW at Langley AFB.

Below right: Streak Eagle easily beat the F-4's climb records, and was faster than an Apollo moonshot to 15,000m.

negate some of the roll rate demanded by small sharp deflection of the stick.

The need to retract the undercarriage into the fuselage led to a rather narrow track of 9ft 0¼in (2.75m) and a wheelbase of 17ft 9½in (5.42m). To have considered another configuration would have incurred an unacceptable weight penalty. The narrow track promised to produce a few problems, and these duly appeared during the flight testing. During crosswind landings the upwind wing would come up, causing the aircraft to tend to weathervane into the wind, and again drift downwind. Holding the nose up on landing only accentuated the problems, and so pilots always tried to get the nosewheel on the ground as quickly as possible.

The causes of the problems were soon identified. The first involved the aileron-rudder interconnect (ARI) system: as the stick was moved laterally (assuming a neutral or aft longitudinal position) rudder movement was initiated in sympathy. So, if the right wing came up and the stick was moved to the right to counteract it, the rudder motion would make the aircraft yaw to the right, and thus aggravate the tendency to weathervane. The second problem occurred with the stick aft, as if to hold the

nose up, when the aircraft systems washed out some lateral control. This had been designed into the controls so as to mimimize lateral deflections of the stick at high angles of attack. As the pilots said, "Rolling out on the runway was not the place to reduce lateral control, particularly in a very lightly wing-loaded fighter with a narrow gear".

In these circumstances, the wind would blow the wing up, and the aircraft would start weathervaning. The normal response by the pilot would be to move the stick into wind, but this did nothing to level the wings, and succeeded only in worsening the yaw into wind, giving the pilot the impression that the aircraft wanted to tip up and over onto the downwind forward quarter. Once the nosewheel was down, the situation improved slightly, but all the characteristics remained to a lesser degree, and the resultant roll-out was described as 'uncomfortable'. The whole problem was exacerbated by the oleo struts on the mainwheels tending to stroke at different times in the roll-out and on a calm day, there could be a 2 or 3deg difference until the up-wing oleo would stroke. A further weak point was the low-gain steering on the nosewheel.

Streak Eagle world time-to-height records

Altitude	Time	Date	Previous time	Margin
3,000m (9,843ft)	27.57sec	Jan 16, 75	34.52sec (F-4B)	20 per cent
6,000m (19,685ft)	39.33sec	Jan 16, 75	48.79sec (F-4B)	19 per cent
9,000m (29,528ft)	48.86sec	Jan 16, 75	61.68sec (F-4B)	21 per cent
12,000m (39,370ft)	59.38sec	Jan 16, 75	77.14sec (F-4B)	23 per cent
15,000m (49,212ft)	77.02sec	Jan 16, 75	114.50sec (F4-B)	33 per cent
20,000m (65,617ft)	122.94sec	Jan 19, 75	169.80sec (MiG-25)	28 per cent
25,000m (82,021ft)	161.02sec	Jan 26, 75	192.60sec (MiG-25)	16 per cent
30,000m (98,425ft)	207.80sec	Feb 1, 75	243.86sec (MiG-25)	15 per cent

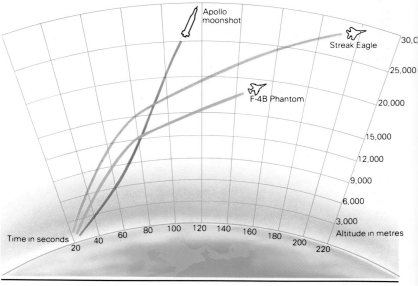

Above: The record-breaking Streak Eagle, 72-0119, over the familiar St Louis skyline.

The full plus or minus 15deg steering should have assisted three-point directional control, but because of the pedal-to-wheel deflection, long, strong legs were required to get positive response.

Considerable effort went into fixing the whole dilemma of cross-wind landing, and the Eagle can now accept such landings in 25–30kt crosswinds. To begin with, the ARI was all but eliminated on touch-down, as it had no essential function on the ground. The mechanical ARI was made to deactivate on sensing wheel-spin on the ground, which was almost instantaneous after touch-down. This effectively eliminated the effects of the ARI, but did retain it in a static condition until the ground checks were made. The CAS ARI was also largely eliminated during the roll-out in a similar fashion.

The mechanical lateral control wash-out with the stick aft was eliminated with the undercarriage down, while the gain was increased on the nosewheel steering so that the response was swifter.

Right: Major D.W. Peterson, who piloted the Streak Eagle to the 15,000m and 25,000m records on January 16 and 26, 1975, by the nose of the specially prepared F-15A.

Above: An early production F-15C refuels from a KC-135A tanker with an example of MCAIR's previous USAF fighter, the F-4, in attendance.

This fix was primarily to improve the taxying qualities of the Eagle, but the benefits to directional control on the runway were welcomed. The main-wheel oleo struts were significantly modified, so as to achieve a greater load stroke quicker on touchdown, while the remainder was taken up at lower speeds: Overall the Eagle had a

Below: One of the virtues of the F-15 is the minimal amount of ground support equipment needed, as illustrated in this view of pre-flight checks at a snow-covered base.

more solid feel following touch-down.

The time and effort expended on this series of fixes have paid dividends for Eagle operations. The 25–30kt cross-wind component allows crab angles of up to 12deg to be used on landing. With the nose held at 12deg pitch, for maximum braking effect, the Eagle's velocity vector is simply held straight down the runway with the rudder until the nose is lowered at about 80kt, where the nose-wheel steering takes over. All the pilot needs do is to fly down the runway using normal aerodynamic control until the nosewheel is lowered to contact with the ground.

The handling qualities of the F-15 Eagle in flight are described as 'excellent' using either the conventional hydro-mechanical system or the CAS to fly the

aircraft. However, several modifications known as Engineering Change Proposals (ECPs) were made, though by autumn 1974 only 36 had been recommended, of which 13 did not involve the Eagle itself. Of the 23 ECPs made to the aircraft, only three are externally visible: the raked wing tips, the dog-tooth stabilator, and the enlarged speed brake.

Early in the test programme MCAIR discovered a buffet and wing-loading problem at certain altitudes. After attempting to solve the problem with wing fences, the solution adopted was to remove 4sq ft (0.37sq m) diagonally from the wing tip to create a raked appearance. The cutting of the dog-tooth into the leading edge of the stabilator was the solution to a flutter

problem discovered in wind tunnel testing. This produced a minor shift in the coefficient of pressure and a change in the moment of inertia sufficient to remove the flutter. The enlargement of the speedbrake from 20 to 31.5sq ft (1.86 to 2.93sq m) allowed the required drag to be produced from lower extension angles, and removed a buffet caused by the original brake at the desired drag configurations (and higher extension angles).

Radio-controlled models
One of the more interesting aspects of the F-15's flight test programme was the use of large glider models of the Eagle, which were dropped from a Boeing B-52 of the NASA Flight Research Center flying at 45,000ft (13,700m) at 175kt. Termed 'remotely piloted research vehicles', these models were built of aluminium and glass fibre to three-eights scale, having a wing span of 16ft 1¼in (4.91m) and a length of 23ft 10¾in (7.28m) and weighing some 2,000lb (907kg). Radio-controlled from the ground, they performed their manoeuvres, deployed a parachute and were recovered in mid-air by a helicopter. Among their tasks were to conduct high angles of attack, stalling and spinning manoeuvres ahead of the live flight tests performed by the eighth development Eagle.

All these ECPs, many of a minor nature, have contributed to making a good flying aircraft into a fighter with excellent handling qualities. Following its first flight, the F-15 Eagle met all its milestones on time with one exception: because of technical problems concerned with F100 engine durability tests, this part was some months late. The test programme's deliberately slow pace was, according to Major General Bellis, able to provide "a significant capability to profit from information derived from the initial and intensive ground-test programme and the current joint Air Force and contractor F-15 flight test activities". Such a pace is totally in keeping with the 'test before fly' and 'fly before buy' attitude adopted by the USAF in their procurement procedure.

Final proof of the new fighter's capabilities was provided by Operational Streak Eagle, a joint USAF/MCAIR operation to break a number of time-to-height world records, previously held

Above: St Louis again forms the background for this February 1979 view of the first F-15C before its delivery to TAC.

by the F-4 Phantom and Soviet MiG-25 Foxbat. The 19th pre-production aircraft (the 17th single-seater) was totally stripped down: radar, cannon, missiles, tail hook, utility hydraulic system, one of the generators, and several actuators were

Below: Three-view drawing of the production F-15C; the upper side profile shows the two-seat F-15D. Without the FAST packs, the F-15C is externally identical to the A.

removed, along with the paint finish, leaving a bare metal aircraft. Between January 16 and February 1, 1975, Majors W. R. Macfarlane, D. W. Peterson and R. Smith established a total of eight new time-to-climb records flying from Grand Forks AFB. The Eagle was some 2,800lb (1,270kg) lighter than a standard production aircraft and only sufficient fuel was carried for the specific flight and return to the airfield.

The Category I and II flight test programme was completed by the delivery of the first production Eagle, a TF-15A (F-15B) to the USAF's Tactical Air Command at Luke AFB on November 14, 1974. On July 1, 1975, the first oper-

ational F-15 Eagle squadron within Tactical Air Command – the 1st Tactical Fighter Wing (TFW) – was formed at Langley AFB, and six months later, on January 9, 1976, this wing received its first aircraft.

As with most aircraft in production, improvements and refinements are constantly emerging. The current production models of the Eagle are the F-15C and F-15D, which replaced the single-seat F-15A and twin-seat F-15B respectively from mid-1979, when a total of 443 had been produced.

Externally there is no major difference between the two sets of Eagles. Internally, the APG-63 radar has been

improved by the addition of a programmable signal processor (PSP), details of which are given in Chapter 4; the fuel capacity has been increased by some 312.5USgal (260Imp gal/1,188 litre); and the aircraft has been modified to accept two conformal Fuel and Sensor Tactical (FAST) packs on either side of the fuselage, enabling the internal fuel capacity to be increased by a further 1,523US gal (1,268Imp gal/5,788 litre) altogether.

The first F-15C Eagle flew on February 27, 1979, with the first F-15D following on June 19, 1979. These versions entered service with the USAF in September 1979, when the 18th TFW at Kadena AB, Okinawa, Japan, received their first squadron. The F-15C/D model represents the baseline from which the F-15 Multi-stage Improvement Programme (MSIP) begins. In addition to the enhancements already noted, the MSIP will take in a new programmable weapon control set, with provision for the AIM-120 AMRAAM, AIM-7M Sparrow and AIM-9M Sidewinder missiles; expanded electronic warfare (EW) equipment, including an enhanced radar warning receiver (RWR), an internal ECM system and the addition of chaff/flare launchers; and expanded communications facilities, Seek Talk, HF radio and provision for JTIDS. Although further versions with enhanced roles are under development, the E Strike Eagle remains the current production standard, and is likely to continue as such in the near term.

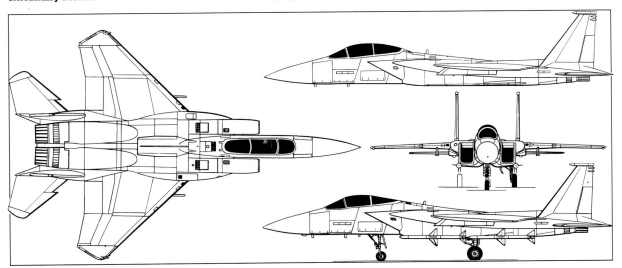

Structure

Thousands of hours of computer studies and wind-tunnel tests went into the F-15's design; billions of dollars have been spent on building the aircraft and its essential systems. Much of this effort and expense was devoted to keeping the basic structure as simple as possible, for maximum survivability in combat and ease of maintenance on the ground. In service, the Eagle may not have quite matched its designers' expectations, but an outstanding safety record and an improving rate of availability demonstrated by Tactical Air Command Squadrons are testimony to its basic soundness.

By mid-1967 the early conception of the FX as a 60,000lb (27,000kg) variable-sweep multirole aircraft had been rejected in favour of a 40,000lb (18,000kg) fixed-wing fighter, and by September 1968, when the Contract Definition RFP was issued, a number of design criteria had been established. Several of these were directly concerned with structural features, while others involving specific operational requirements imposed their own indirect constraints on the physical details of the resulting aircraft.

Since the Eagle was to be used for air combat, a combination of high thrust-to-weight ratio and low wing loading were stipulated, within the overall gross weight limit of 40,000lb. A high degree of survivability was called for in structure and subsystems, involving com-

prehensive testing of components used, along with extended component life and reduced maintenance requirements. And the single crew member, who would be provided with a comprehensive suite of automated avionics to enable him to carry out his mission unaided, was to be given all-round visibility.

Computer evaluations

During the preliminary FX design studies – which ultimately involved some 2,500,000 man-hours and resulted in 37,500 pages of documentation – MCAIR engineers carried out extensive computer evaluations comparing the weights of hypothetical aircraft of given cost resulting from a variety of basic design features. For example, if a two-

man, all weather avionic suite were installed, the gross weight of the aircraft was calculated as 46,000lb (20,865kg), while single-seat, clear-air avionics required a gross weight of only 31,500lb (14,290kg). Similar comparisons were made between theoretical aircraft with varying degrees of energy manoeuvrability and maximum speeds ranging from Mach 0.8 sea-level dash to a sustained Mach 2.7.

The parameters for the structural component of the computer evaluation were represented by an upper limit of 8g with 100 per cent fuel and a lower capacity for 6.5g with only 60 per cent fuel, giving respective gross weight figures of 41,500lb (18,825kg) and 38,000lb (17,235kg). The actual design limitation of 7.33g was achieved within

Above: Viewed from above, the contours of the Eagle fuselage clearly show the central pod and twin boom structural configuration.

the specified 40,000lb weight limit.

Considering that the external dimensions of the F-15 are marginally greater than those of the F-4 Phantom, it is remarkable that the maximum takeoff weight of the newer fighter should be nearly 6,000lb (2,700kg) less than that of its predecessor. Part of this reduction is a result of the lighter internal fuel load, which with the Eagle's more efficient

Below: June 1981, and among the F-15Cs and Ds on the MCAIR assembly line is the distinctive tail of an F/A-18 Hornet.

engines still gives a longer range, and the rest is accounted for by the higher percentage of titanium and advanced composite materials used in its construction. As originally designed the F-4 was built with more titanium than any previous fighter, but even in its later versions this only amounted to some 9 per cent by weight of the total structure, whereas the Eagle airframe includes 25.8 per cent titanium, only 5.5 per cent steel and 37.3 per cent aluminium.

The titanium is largely concentrated around the engines and the inboard sections of the wings. The fuselage itself is of conventional semi-monocoque construction, and its pod and twin boom configuration is apparent in the contours of the upper surfaces. The frames of the central pod and of the air intakes on either side and their skin are of machined aluminium, as is the front wing spar, but the three main wing spars and the bulkheads connecting them and the frames of the engine pods are of titanium, also machined. Aft of the forward main wing spar the fuselage skin is also of titanium, and the same metal forms the cantilever booms outboard of each engine which carry the twin fins and horizontal stabilators, the stabilator attachments and the spars of the fins.

Titanium strength
There are several advantages to the use of titanium in these areas. The resulting structure is strong enough to transmit the control forces from the tail surfaces, and the strength of the twin engine bays, with titanium firewalls between them, reduces the risk of a fire or explosion in one engine damaging or incapacitating the other. Similarly, the titanium skin of the inboard underwing skin covers the integral fuel tanks, which are filled with sealant foam injected through channels in the joints between the spars and the aluminium upper skin. Moreover, the titanium wing spars and supporting bulkheads are strong enough to allow the aircraft to keep flying with any one spar in each wing completely severed.

Further weight savings are achieved by the use of composite materials over honeycomb cores for the wing flaps and ailerons and the tailfins and stabilators. The central sections of the tail surfaces, where the titanium structural members form torque boxes, as well as the rudders, are covered with boron composite skins with aluminium and Nomex honeycomb between, while aluminium skins cover the honeycomb flaps and ailerons and the raked outboard sections of the wings. The speedbrake, similarly, has a graphite composite skin

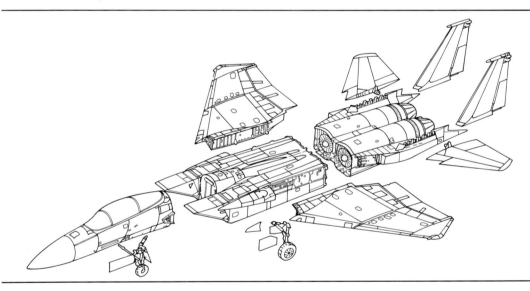

Below: The bubble canopy is the only external feature of the Eagle to interrupt the aerodynamically clean contours of the airframe.

Above: Major structural assemblies of the F-15 airframe. Tail surfaces and wings are interchangeable between aircraft.

Top: Automatic drilling machines in action, forming the attachment holes around the edges of the acrylic windshield destined for an F-15.

over a core of aluminium honeycomb.
The wing torque box assembly is based on the three titanium spars, extended by aft and outer spars of aluminium, and the wings are attached to the titanium bulkheads by pin joints which allow interchangeability between aircraft. The flaps and ailerons are similarly interchangeable, as are the windshield, canopy, speedbrake and fibreglass nose radome, while the vertical stabilizers, rudders and stabilators are interchangeable between right and left sides of the aircraft.

The aircraft is constructed in stages, with individual assemblies built up separately and progressively integrated. The main assemblies are the rear, centre and forward fuselage sections, the vertical and horizontal stabilizers and the wings, and various techniques are used in forming the multitude of sub-assemblies involved.

Sheet aluminium parts are drilled and routed automatically, with more sub-

stantial aluminium sections formed in a two-storey press exerting a pressure of 7,000 tons. Titanium forgings are also finished by computer-controlled milling machinery, and titanium sheets are formed in a furnace at temperatures of 1,650deg F (900deg C) and pressures of 250psi (45kg/sq cm). Graphite skins for the speedbrake are shaped by a high-speed laser cutter, while the honeycomb core is mechanically carved to match. Other automated processes include computerized pipe bending and robot drilling of the attachment holes for the acrylic windshield.

Meanwhile, sheets of the boron epoxy composites used for the tail surfaces are assembled by hand before being bonded by high-pressure steam in an autoclave. Metallic parts are spray-painted individually to protect them from corrosion, and critical sub-assemblies such as the engine intake ducts and wing leading edges are drilled and riveted by hand. Another job for

Above: Ground crew work under floodlights to prepare a 49th TFW F-15A for a sortie during Red Flag exercises at Nellis AFB in 1980.

Below: A ground crewman sporting an 'Eagle Keeper' badge takes advantage of the ready accessibility of the F-15's internal systems.

Left: An F-15B from Nellis AFB's 57th TTW in its hangar for routine servicing during a Red Flag desert warfare simulation.

skilled technicians is the assembly of the 290 bundles of electric wiring: these are installed during the build-up of the fuselage sections, as are fuel and hydraulic systems.

The final assembly stage brings together the three fuselage sections and the wings, and the final connections of all the power, control and environmental systems. These are checked before the final stage, which involves the installation of engines and avionics, and after a final checkout the aircraft is flown by a company test pilot to ensure that all systems are functioning.

As well as the parts fabricated in MCAIR's own plant, the F-15 involves the use of parts supplied by some 1,200 sub-contractors whose value is approximately half that of the total aircraft cost. By 1980 total cost of the F-15 programme had amounted to $16.58 billion, giving an average unit cost for the 749 aircraft involved of rather more than $22 million (in FY 1980 dollars). The 60 F-15s sold to Saudi Arabia cost a total of $1.92 billion, while Israel's first 40 Eagles were worth $907 million.

Design rationale

The wing configuration is relatively straightforward, though it was only selected after several hundred had been analysed and more than 100 were tested in almost a year of solid wind-tunnel trials. Ultimately, variable camber, with movable surfaces on both leading and trailing edges as used on the F-16, was rejected, since an alternative design with a fixed leading edge employing conical camber was found to offer only slightly higher supersonic drag and marginally reduced subsonic performance, both of which were more than offset by advantages in terms of

weight, simplicity of manufacture and ease of maintenance. The chosen design has a straight leading edge swept at an angle of 45deg, an aspect ratio of 3, zero incidence and 1deg of anhedral.

As outlined in the previous chapter the wingtips were modified from their original shape, a total of 3sq ft (0.28sq m) being removed from each tip to reduce excessive lift at the wingtips and consequent severe buffeting at high g loads and high subsonic speed at altitudes of 30,000ft (9,000m) or so. Another change made as a result of flight testing was the increase in chord of the outer portions of the horizontal stabilators, creating the characteristic notched leading edges of these surfaces, to eliminate flutter. At the same time, the original speedbrake was increased in size by more than 50 per cent, again to eliminate a buffet problem.

Overall, the F-15 profile is designed to minimize wave drag at high speeds, with the single exception of the bubble canopy, which is a feature of both the Eagle and the USAF's other new fighter, the F-16 Fighting Falcon. The aerodynamic penalties imposed by such a canopy are regarded as unavoidable if the pilot is to be enabled to perform his air superiority mission effectively.

The canopy itself is a single transparency with only one transverse frame, hinged at the rear and counterbalanced by a single strut, with a second strut immediately behind the pilot's seat for emergency jettison of the canopy. The size of the canopy is the only external difference between single-seat and two-seat Eagles: the latter have a longer canopy to accommodate the rear crew member, while the single-seater has an avionics bay behind the pilot's seat. This

Right: For maximum maintainability the F-15 airframe features a total of 570sq ft (53sq m) of access doors and panels, as demonstrated here.

houses several avionics 'black boxes' associated with the Tactical Electronic Warfare System in current models but is largely empty and offers ample room for additional avionic equipment; it is unpressurized, and when the canopy is lowered it is isolated by an integral seal.

Directly behind the avionics bay are the fuselage fuel tanks, and between the forward and aft tanks, below the speedbrake, are the ammunition drum and feed system for the M61 cannon. The gun itself is housed in the starboard wing root fairing, while the equivalent fairing at the port wing root contains the flight refuelling receptacle.

The only examples of variable geometry in the F-15's structure are the engine air inlets on either side of the forward fuselage. Because the aircraft was designed to be flown at high angles

Right: Extending steps built into the fuselage side of the F-15 can be used instead of the normal ladder for cockpit access.

of attack in combat, the intakes are able to 'nod' up or down to keep the aperture facing directly into the airflow in order to maintain an adequate supply of air to the engines. The intakes are pivoted at their lower edge and adjusted to angles of 4deg above or 11deg below the horizontal by hydraulic actuators controlled by the air data computer. The intake angle can also be adjusted to prevent more air than necessary being

Below: An 8th TFW Eagle is given an automatic wash at Kwang Ju air base in Korea, during Exercise Cope North in June 1982.

Left: The remarkable slenderness of
the F-15's twin vertical tails is
achieved by the use of boron
composite skins on honeycomb cores.

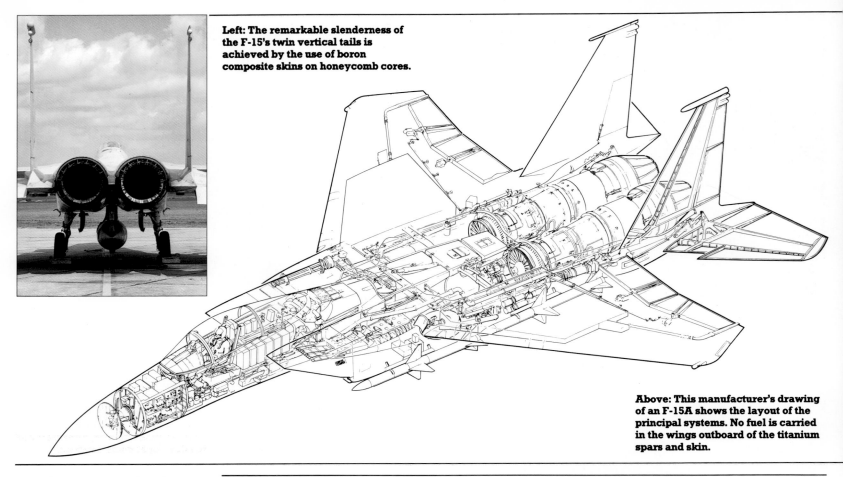

Above: This manufacturer's drawing
of an F-15A shows the layout of the
principal systems. No fuel is carried
in the wings outboard of the titanium
spars and skin.

Right: Wing leading and trailing edge
tanks, and additional tanks in the
centre fuselage, allow the F-15C to
carry an extra 1,855lb (880kg) of fuel
internally.

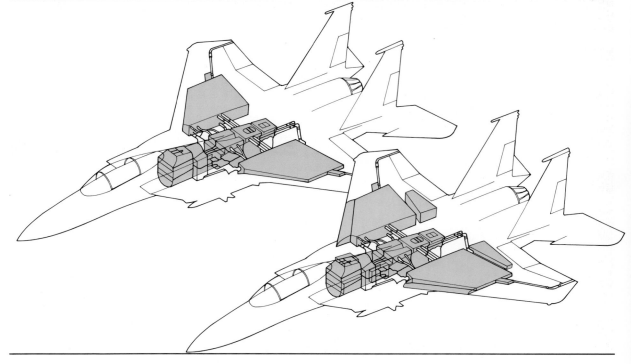

taken in, and the intake surfaces have a
further function in providing additional
manoeuvring control in a similar
manner to canard foreplanes. At supersonic speeds their effectiveness is
almost a third as great as that of the
stabilators, whose size and weight were
enabled to be reduced in consequence.
Immediately aft of the intakes themselves are twin mechanically linked
ramps to control the shockwaves created in the incoming air.

Of course, structural and aerodynamic efficiency are prerequisites in
any aircraft, but a combat aircraft must
also be designed to survive the inevitable wear and tear of battle. In this
respect the Eagle's main assets are
reckoned to be its overall superiority in
such areas as pilot visibility, radar detection ability, weapons systems and
performance, but a number of less obvious safety features were included,
largely as a result of experience with
the F-4 in Vietnam, in addition to the
basic structural elements described
above.

Hydraulics and fuel

As well as twin engines, the F-15 has
three separate hydraulic systems which
can detect and isolate leaks in their
associated subsystems and each of
which is capable of sustaining the flight
control system on its own. Similarly, twin
electrical systems, powered by 40/
50kVA AC generators operating
through DC conversion units, are capable independently of fulfilling the aircraft's power requirements, while a
standby hydraulic generator is fitted to
supply critical systems in an emergency. Either the electronic control
augmentation system or the hydro-
mechanical system can provide inde-
pendent flight control should one or the
other be put out of action.

Fuel supply is another vital system,
and this too is designed to cope with

various emergencies. The fuel tanks
themselves are filled with foam, and the
fuel lines are also self-sealing. In the
event of electrical failure, the fuel
system will continue to operate on
standby power, providing fuel flow adequate for engine operation up to the
mid-afterburner range.

Additional protection against fire is
provided in the form of a fire-suppres-
sion system. A pressurized bottle con-
taining a non-corrosive agent is located
between the engine bay firewalls, with
three nozzles able to release the agent
into either engine or in the space between them. The provision of this fire
suppression system is one of the direct
results of the F-4's experiences in Vietnam, and the Eagle is one of the few
fighter aircraft to be fitted with such
equipment. The physical separation of
the fuel tanks from the engine bays is
another precaution against fire.

Another aspect of the F-15 design

intended to maximize its efficiency in
combat is the emphasis placed on ease
of maintenance and relative independence of ground support equipment. The
airframe features a total of 570sq ft
(53sq m) of access panels, allowing most
routine maintenance to be carried out
without the use of work stands. The
overall simplicity of the airframe is a
major factor in reducing maintenance
requirements: by comparison with the
F-4E, the F-15 has only 202 lubrication
points against the earlier fighter's 510;
seven hydraulic filters, all of which are
interchangeable and which incorporate
visual indications of the need for replacement, are used in place of the
Phantom's 21; plumbing connections in
the fuel system are reduced from 281 to
only 97, and the interchangeable fuel
pumps are of a plug-in design that
allows them to be changed in only 30
minutes; and whereas the F-4E had 905
potted electrical connectors, whose

waterproof compounds were subject to
deterioration, the F-15 has a total of 809
silicone grommet connectors to avoid
this problem.

The digital avionic systems used also
help to reduce the maintenance workload. No routine servicing is needed for
these: they are line replaceable units
(LRUs), readily removable in the event
of failure for testing and repair, and
faulty units can be located by means of a
built-in test panel located in the nosewheel well. The cockpit is also equipped with a caution light panel to indicate system failures and a built-in test
(BIT) panel for the avionic systems. To
reduce dependence on support equipment the secondary power system
which provides power for the jet fuel
starters can also power the electrical
and hydraulic systems for up to 30
minutes while maintenance work is carried out or for munitions loading.

Power for the electrical generators

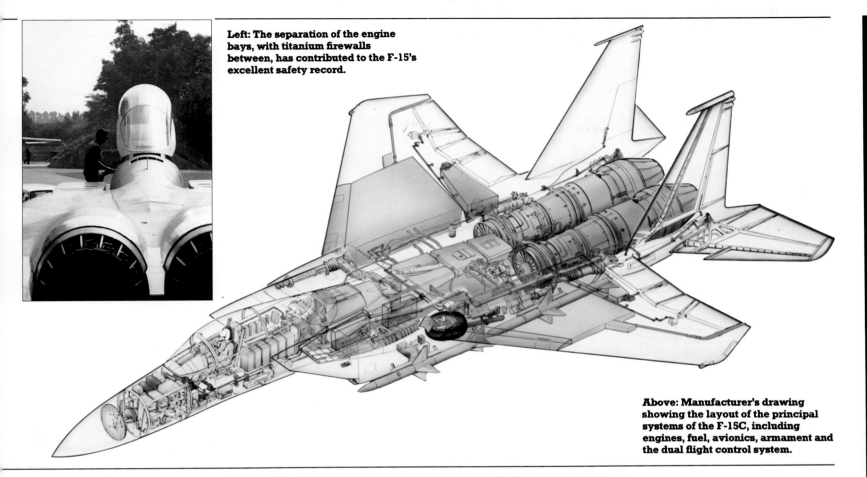

Left: The separation of the engine bays, with titanium firewalls between, has contributed to the F-15's excellent safety record.

Above: Manufacturer's drawing showing the layout of the principal systems of the F-15C, including engines, fuel, avionics, armament and the dual flight control system.

Left: The increased maximum takeoff weight of the F-15C necessitated strengthening of the undercarriage, tyres and brakes. A 36th TFW F-15C shows its landing gear.

and hydraulic pumps is derived from the main powerplant via an aircraft-mounted accessory drive shaft. Being mounted on the aircraft rather than in the engines themselves, this allows the engines to be interchangeable between right and left. To simplify engine maintenance, each F100 turbofan can be broken down into five main modules, consisting of the inlet and fan, the gearbox, the fan drive, the engine core and the augmentor (afterburner) duct and nozzle. Any of these can be removed for maintenance while another module is substituted, so that a malfunction in one part of the engine does not put the whole powerplant out of action while it is repaired. The engines themselves are removed quickly and easily, sliding out of the rear of the bays on integral rails onto a cart with matching rails, and with only ten disconnections necessary before removal. Complete engine change can be accomplished well

inside the 30 minutes minimum specified by the Air Force. And with the engines installed internal inspections can be carried out by means of 12 borescope ports on each engine.

The overall intention behind this careful planning of maintenance features was to meet a USAF requirement for a maximum of 11.3 maintenance man-hours per flight hour (MMH/FH), just under half the requirement of the F-4E, which would allow a 15 per cent reduction in the number of maintenance personnel. In practice, however, this figure has not been achieved: reduced to 19 by the time type testing was complete, the MMH/FH figure had risen to around 35 by 1980. One reason for this was the continuing difficulties with the engine, as described in the following chapter, with the powerplant absorbing nearly half the total maintenance effort. Another was the continuing shortage of skilled maintenance personnel, partly

as a result of the comparatively low rates of pay which such personnel received compared with those available elsewhere.

An equally serious problem was the difficulties encountered with the Avionics Intermediate Shops (AIS), the second of three stages in the F-15 planned maintenance programme. The first level of maintenance takes place on the flight line, where routine servicing is carried out and line-replaceable units are removed as necessary. The second stage, of which the AIS forms part, is the repair on base of faulty units, and the third is depot repair at Air Logistic Centres.

Among the problems encountered with this system was the difficulty of repairing LRUs on base, with nearly half of the avionics LRUs being returned to the depots. This in turn meant that stocks of replacement LRUs were quickly exhausted. The principal reason

for this state of affairs was the repeated failures of the automated testing stations which comprise the AIS, to the extent that only half the AISs were operational at any time. And the problem was compounded by the failure of the AIS to agree in its diagnosis of the fault with the indications of the on-board BIT equipment in a large percentage of cases.

The natural consequence of all these problems was a high rate of cannibalization, with parts being removed from one aircraft to keep another flying. Consequently, by 1979 availability of the F-15 within TAC had fallen to an alarmingly low 56 per cent. On the other hand, a high degree of readiness has been achieved on numerous exercises, and the Eagle has compiled an impressive safety record. Overall availability by 1983 had risen to more than 65 per cent, and an official inspection of the 1st TFW at Langley in August 1982 found a record 98.6 per cent of aircraft operational: this compares with an availability of only 35 per cent of the same wing's aircraft three years earlier.

New model

Meanwhile, the original F-15A and B production models were replaced by the improved C and D from June 1979. Again, external differences between the single and two-seat version are limited to the slightly longer canopy fitted to the F-15D, and the principal external distinguishing feature of the later models is the provision of the FAST Pack conformal fuel tanks.

The FAST Packs are attached to the outside of each air intake, and can carry an additional 4,875lb (2,211kg) of fuel. Alternatively, sensors such as reconnaissance cameras or infra-red equipment, radar warning receivers and jammers, laser designators and low-light TV cameras, or a combination of such sensors with reduced quantities of fuel, can be carried. When the FAST Packs are fitted, the four Sparrow missiles are mounted on their corners, and bombs or air-to-surface missiles weighing up to 4,400lb (1,995kg) can be carried as an alternative.

Internal fuel capacity of the F-15C/D is also increased by 1,855lb (880kg), with additional tanks located in the centre fuselage and in the wing leading and trailing edges. The effect of this increased fuel capacity is to raise the potential gross weight of the F-15C to 68,000lb (30,845kg) with full internal fuel, FAST Packs and three external tanks. As a result, tyres, wheels and brakes have been strengthened to cope with the increased weight.

The FAST Packs alone carry slightly less fuel than the normal three external tanks, but allow the aircraft to be flown at considerably higher speeds. Compared with the clean configuration, an F-15 equipped with FAST Packs experiences only slightly increased profile drag at subsonic speeds, and compared with the standard external tanks the conformal tanks contribute only a fraction of the former's drag at supersonic speeds. Ferry range with the increased internal fuel, FAST Packs and wing and fuselage tanks is increased to 3,450 miles (5,560km). According to the manufacturers, unused space in the outboard sections of the wings could accommodate a further 900lb (400kg) of fuel.

The manufacturers have also proposed to exploit currently unused internal space, particularly in the rear of the cockpit and in the tail booms, to equip the Eagle for other roles. As much as 56cu ft (1.6cu m) of growth space is claimed to be available for additional avionics, and in the Strike Eagle/Advanced Fighter Capability Demonstrator version of what was originally the second development F-15B the company has already installed a weapon system operator's station in the rear cockpit. Other roles envisaged for developed versions of the Eagle include specialized reconnaissance and Wild Weasel defence suppression models.

The latter would be intended to replace, respectively, the RF-4C and F-4G versions of the aging Phantom, though USAF evaluations during 1983 were concentrated on the dual-role fighter/ground attack role, for which the Eagle was subjected to comparative trials with the F-16XL.

McDonnell Douglas F-15C Eagle cutaway drawing key

1 Tailplane honeycomb construction
2 Boron fibre skin panel
3 Tailplane spars
4 All-moving tailplane pivot fixing.
5 Leading edge dog-tooth
6 Low-voltage formation lighting strip
7 Fin root attachment frames
8 Rudder hydraulic rotary actuator
9 Rudder honeycomb construction
10 Fin spar construction
11 Boron fibre skin panel
12 Anti-collision light
13 Electronic countermeasures aerials (ECM)
14 Variable area afterburner exhaust nozzles
15 Nozzle sealing flaps

16 Fueldraulic nozzle actuators
17 Afterburner duct
18 Engine bay titanium ring frames
19 Rear engine mounting frame
20 Engine bay titanium frame and stringer construction
21 Titanium skin panelling
22 Port tailplane hydraulic actuator
23 Tailplane hinge arm
24 Port rudder
25 Tailboom fairing
26 ECM aerial
27 Port tailplane
28 Tail navigation light
29 ECM aerial
30 Radar warning aerials
31 Boron fibre skin panelling
32 Fin leading edge
33 Port air system equipment bay
34 Forward engine mounting
35 Engine mounting frame
36 Bleed air system ducting
37 Engine support link
38 Engine bay fireproof bulkhead
39 Pratt & Whitney F100-PW-100 afterburning turbofan engine
40 Starboard air system equipment bay
41 Engine bleed air primary heat exchanger
42 Heat exchanger ventral exhaust duct
43 Retractable runway arrester hook
44 Wing trailing edge fuel tank
45 Flap hydraulic jack
46 Starboard plain flap
47 Flap and aileron honeycomb panel construction
48 Starboard aileron
49 Aileron hydraulic actuator
50 Fuel jettison pipe
51 Aluminium honeycomb wing tip fairing
52 Low-voltage formation lighting
53 Starboard navigation light
54 ECM aerial
55 Westinghouse ECM equipment pod
56 Outboard wing stores pylon
57 Pylon attachment spigot
58 Cambered leading edge ribs
59 Front spar
60 Machined wing skin/stringer panels

61 Outboard pylon fixing
62 HF flush aerial
63 Leading edge fuel tank
64 Inboard pylon fixing
65 Wing rib construction
66 Starboard wing integral fuel tank, total internal fuel load, 13,455lb (6103kg)
67 Wing root rib support struts
68 Titanium wing spars
69 Wing spar/fuselage attachment pin joints
70 Machined fuselage main bulkheads
71 Wing/fuselage fuel tank interconnections
72 Airframe mounted engine accessory gearbox
73 Standby hydraulic generator
74 Jet fuel starter (JFS)/auxiliary power unit (APU)
75 Engine intake compressor face

76 Cooling system intake bleed air spill duct
77 Port wing trailing edge fuel tank
78 Port plain flap
79 Flap hydraulic jack
80 Aileron control rod
81 Aileron hydraulic actuator

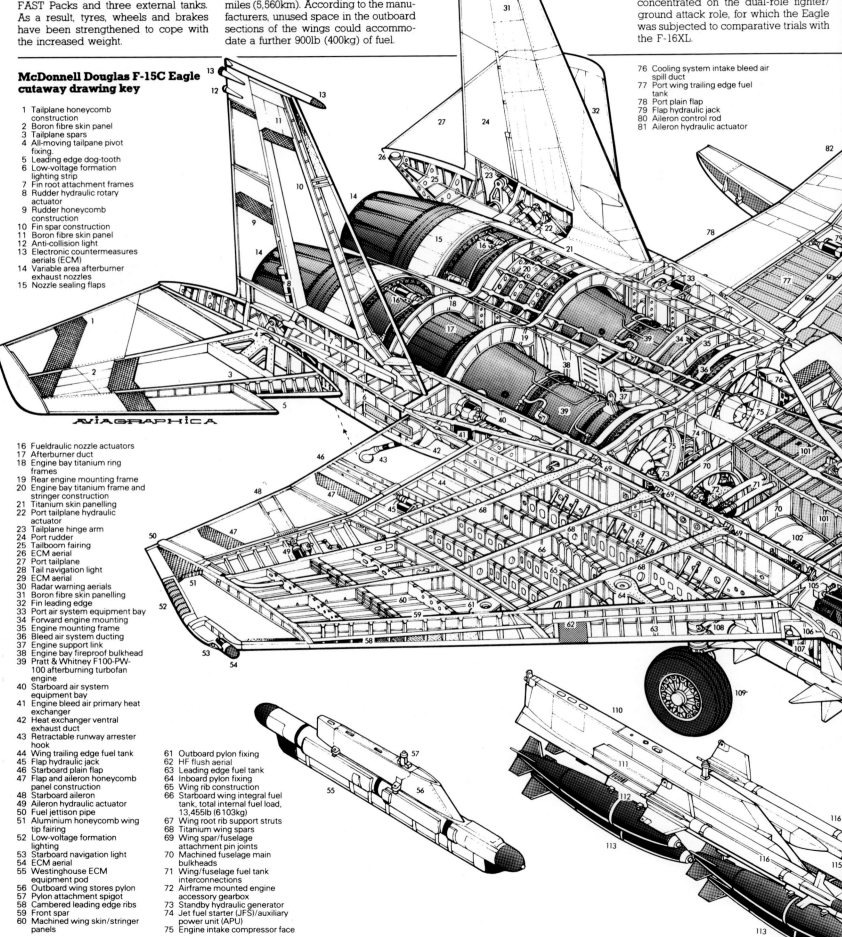

AVIAGRAPHICA

82 Port aileron
83 Fuel jettison pipe
84 Wing tip fairing
85 Low-voltage formation lighting
86 Port navigation light

102 Intake ducting
103 Ammunition feed chute
104 M61A-1 Vulcan 20mm cannon
105 Hydraulic rotary cannon drive unit
106 Starboard anti-collision light
107 Ventral main undercarriage wheel bay
108 Main undercarriage leg strut
109 Starboard mainwheel
110 Inboard stores pylon
111 Air-to-air missile adaptor
112 Bomb rack
113 Mk 82 low drag 500lb (227kg) HE bombs
114 Bomb triple ejector rack
115 Missile launch rail
116 AIM-9L Sidewinder air-to-air missile

121 Central ammunition drum, 940 rounds
122 Airbrake hinges
123 Forward fuselage fuel tanks
124 UHF aerial
125 Intake duct bleed air louvres
126 Intake bypass air spill duct
127 Variable area intake ramp hydraulic actuator
128 Air conditioning system cooling air exhaust duct
129 Canopy hinge point
130 Air conditioning plant
131 Intake incidence control jack
132 Intake duct variable area ramp doors
133 Intake pivot fixing
134 Starboard engine air intake
135 Nosewheel leg door
136 Nose undercarriage leg strut
137 Nosewheel
138 Landing/taxying lamps
139 Nosewheel retraction strut
140 Rear underfloor equipment bay
141 Tactical electronic warfare system (TEWS) racks

147 Cockpit aft decking
148 Canopy arch
149 Port intake external compression lip
150 Fuel and sensor tactical (FAST) pack, conformal fuel pallet, capacity 5,000lb (2268kg)
151 600US gal (2270 litre) external fuel tank
152 Cockpit canopy cover
153 Ejection seat headrest
154 Seat safety handle/arming lever
155 Canopy emergency jettison linkage
156 Ejection seat launch rails
157 Safety harness
158 McDonnell Douglas ACES II "zero-zero" ejection seat
159 Cockpit sloping bulkhead
160 Pilots side console panel
161 Air conditioning ducting
162 Forward underfloor equipment bay, built-in test equipment (BITE) and liquid oxygen converter
163 Low-voltage formation lighting strip
164 Port side retractable boarding ladder
165 TACAN aerial
166 Angle of attack probe
167 Rudder pedals

Above left: With its paint removed the Streak Eagle clearly shows the boron composite skins of its rudder and vertical stabilizer.

Above: Control surfaces of the F-15 are limited to wing flaps and ailerons, twin rudders and all-moving horizontal stabilators.

ACES II ejection seat

A Environmental sensor pitots
B Recovery parachute container
C FLCS data recorder
D Recovery parachute risers
E Emergency oxygen bottle
F Emergency oxygen pressure gauge
G Inertia reel knob
H Ejection control safety lever
I Radio beacon switch
J Survival kit (under seat pan)
K Ejection handle
L Restraint emergency release handle
M Lap belt and survival kit attachment
N Emergency oxygen fitting

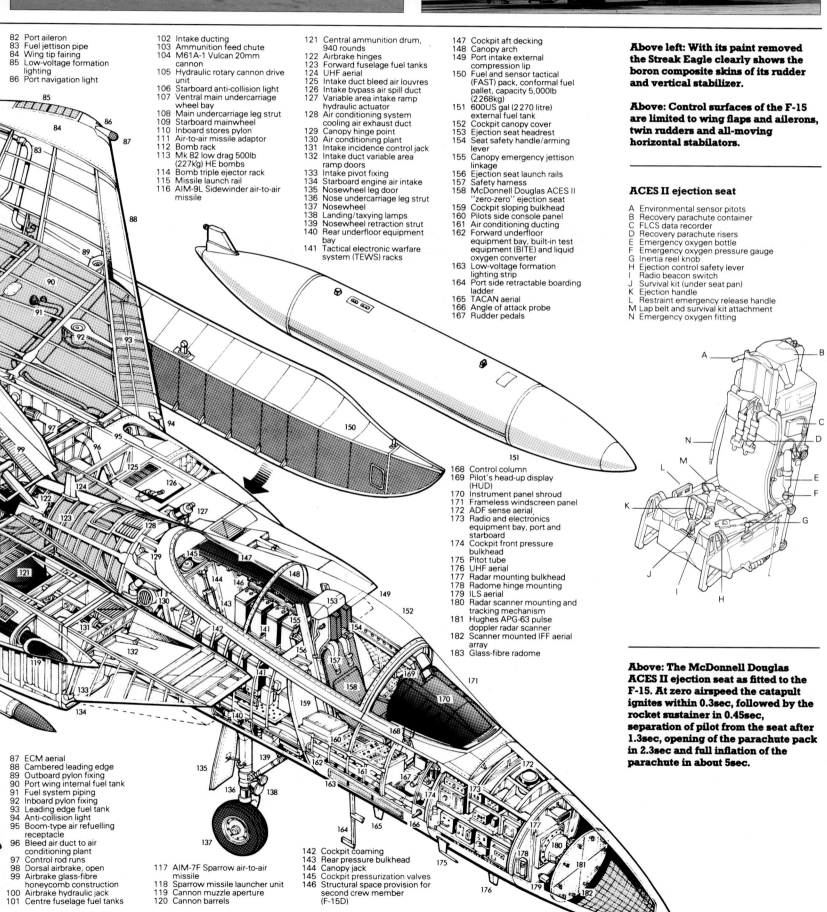

87 ECM aerial
88 Cambered leading edge
89 Outboard pylon fixing
90 Port wing internal fuel tank
91 Fuel system piping
92 Inboard pylon fixing
93 Leading edge fuel tank
94 Anti-collision light
95 Boom-type air refuelling receptacle
96 Bleed air duct to air conditioning plant
97 Control rod runs
98 Dorsal airbrake, open
99 Airbrake glass-fibre honeycomb construction
100 Airbrake hydraulic jack
101 Centre fuselage fuel tanks

117 AIM-7F Sparrow air-to-air missile
118 Sparrow missile launcher unit
119 Cannon muzzle aperture
120 Cannon barrels

142 Cockpit coaming
143 Rear pressure bulkhead
144 Canopy jack
145 Cockpit pressurization valves
146 Structural space provision for second crew member (F-15D)

168 Control column
169 Pilot's head-up display (HUD)
170 Instrument panel shroud
171 Frameless windscreen panel
172 ADF sense aerial
173 Radio and electronics equipment bay, port and starboard
174 Cockpit front pressure bulkhead
175 Pitot tube
176 UHF aerial
177 Radar mounting bulkhead
178 Radome hinge mounting
179 ILS aerial
180 Radar scanner mounting and tracking mechanism
181 Hughes APG-63 pulse doppler radar scanner
182 Scanner mounted IFF aerial array
183 Glass-fibre radome

Above: The McDonnell Douglas ACES II ejection seat as fitted to the F-15. At zero airspeed the catapult ignites within 0.3sec, followed by the rocket sustainer in 0.45sec, separation of pilot from the seat after 1.3sec, opening of the parachute pack in 2.3sec and full inflation of the parachute in about 5sec.

Powerplant

A fighter pilot needs as much power as he can get, so a fighter designed to be the best in the world needs most power of all. The thrust demanded for the F-15 pushed US engine technology to its limits, and the F100 turbofan has had its share of problems. But when a principal source of engine wear turns out to be pilots flying aircraft in ways that were never possible before the manufacturers have reason to congratulate themselves. Meanwhile, the F100 has formed the basis for newer powerplants, and there are new engines under development as possible replacements in a new generation of Eagles.

Development of the Pratt & Whitney F100 turbofan started in August 1968, when the USAF awarded development contracts to P&W and General Electric for engines suitable for use in the planned FX fighter. In view of the high thrust-to-weight ratio planned for the new fighter, the resulting engines would have to push the technology of the time to its limits. P&W faced the daunting task of developing a powerplant producing 25 per cent more thrust per unit of weight than the then-current TF30 turbofan used in the F-111, and twice that of the J75 turbojet used in the F-105 Thunderchief and F-106 Delta Dart.

Both companies built and ran demonstration engines whose light weight, high thrust and low fuel consumption were well in advance of previous designs. The P&W engine was selected by the USAF for further development, contracts being awarded in 1970. Two versions were originally planned – the F100 for the USAF and the F401, intended to power later models of the US Navy's

Below: An F-15A of the 32nd TFS, its engines in full afterburner, accelerates down the runway at Camp New Amsterdam.

F-14 Tomcat, but the latter was cancelled when the USN was ordered by the Department of Defense to cut back the size of the planned F-14 fleet.

The F100 is an axial-flow turbofan with a bypass ratio of 0.7:1. It has two shafts – one carrying a three-stage fan driven by a two-stage turbine, the other carrying the ten-stage main compressor and its two-stage turbine. The completed engine is 191in (4.85m) long and 34.8in (0.88m) in diameter at the inlet, and weighs 3,068lb (1,391kg).

Powder metallurgy

New technologies used in the F100 included powder metallurgy. Instead of forming some metal components in the traditional manner, P&W reduced the raw material to a powder. This could be heated and formed under high pressure to create engine components better able to tolerate the high temperatures planned for the F100 core.

Operating temperature of the F100 turbine was far above that of earlier engines. Successful turbojets of earlier vintage, such as the GE F85 which powers the F-5E, or the GE J79 used in the F-4 and F-104, had turbine inlet temperatures of around 1,800deg F

(982deg C). P&W had achieved figures of just over 2,000deg F (1,093 deg C) in the TF30 turbofan, but to meet the demanding requirements of the F100 specification involved temperatures of 2,565deg F (1,407deg C).

Use of such advanced technology resulted in an engine capable of providing the high levels of thrust required. Maximum thrust is normally described as being 'in the 15,000lb (6,800kg) thrust class' when running without afterburner, and 'in the 25,000lb (11,340kg) class' when full augmentation is selected.

Normal dry (non-afterburning) rating is 12,420lb (5,634kg), rising to a maximum of 14,670lb (6,654kg) at full Military Intermediate rating – the maximum attainable without afterburning. Specific fuel consumption (sfc) – the amount of thrust produced for each pound of fuel burned per hour – is 0.69 at normal rating and 0.71 at Military Intermediate. At full afterburning power, the F100 develops 23,830lb (10,809kg) of thrust at an sfc of 2.17. At this rating, the engine swallows an impressive 860lb (390kg) of fuel per minute.

By the time the F-15 was ready for its first flight in July 1972, the F100 had

Above: The convergent/divergent nozzles of an F-15's twin Pratt & Whitney F100s fully open in the afterburning position.

completed most of its test programme, meeting 23 out of 24 critical 'project milestones'. Between February and October of the following year, a series of turbine failures dogged attempts to complete the 150-hour running trial which formed part of the formal Qualification test. The latter was the most punishing series of tests to which any US military jet engine had ever been subjected, according to P&W. It included 30 hours of running at a simulated speed of Mach 2.3, and 38 hours of running at a simulated Mach 1.6.

Following completion of this test, the F100 was subjected to a further series of intensive trials, including 150 hours of running at over-temperature conditions, and a long series of accelerated Mission Tests. Conducted on the ground, but designed to simulate the stresses of operational service, these were intended to build up running time and detect potential problems. The latter were not serious enough to delay the start of production. The powerplant is

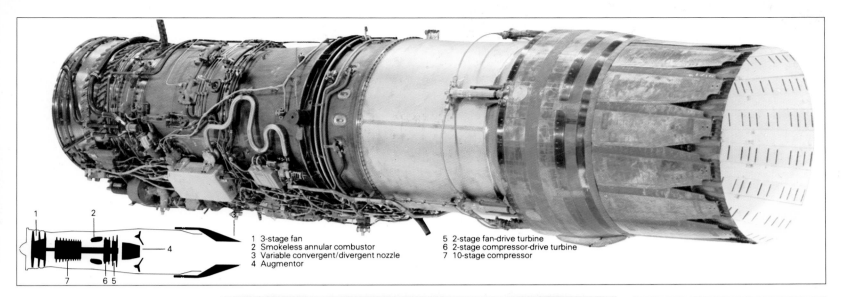

1 3-stage fan
2 Smokeless annular combustor
3 Variable convergent/divergent nozzle
4 Augmentor
5 2-stage fan-drive turbine
6 2-stage compressor-drive turbine
7 10-stage compressor

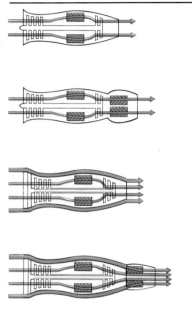

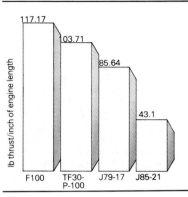

Above: Comparison of the thrust available per inch of engine length from powerplants of (left to right) the F-15, F-111F, F-4E and F-5E.

Top: Exterior of the F100, with diagram of its principal components. Nearly half its length is occupied by the augmentor chamber and nozzle.

Above: Evolution of the afterburning turbofan (bottom) via the turbojet (top), afterburning turbojet and straight turbofan.

designated F100-PW-100 by the company and JTF22A-25A by the USAF.

USAF hopes that the F100 would be a fully-reliable powerplant by the time the F-15 entered service were confounded by a series of technical and operational problems. Strikes at two major subcontractors delayed the delivery of engines, while service experience showed that the F100 was wearing out twice as fast as had been predicted. By the end of 1979 the USAF was being forced to accept engineless F-15 airframes, and by the spring of the following year around 30 were in storage. A massive effort by P&W brought the delivery situation under control, but for a long time the F-15 fleet remained short of engines.

A modification introduced into the fuel pump of the F100 created problems for the F-15 early in its career. In cruising flight, cavitation could begin in the pump, badly damaging the pump vanes. The solution adopted was simple – revert to the original design. In the case of the single-engined F-16 Fighting Falcon, which also uses the F100, a pump failure would be more serious, so Sunstrand developed an alternative dual-element pump for this aircraft. This runs at a lower speed, and should one section fail, the other can continue to deliver fuel at a lower rate.

The electronic engine control unit uses the fuel as a coolant. This technique for obtaining 'free' cooling led to problems when the F-15 first entered

Above: Factory inspection of an F100. The engine weighs 3,068lb (1,391kg) and has a thrust-to-weight ratio of nearly 8:1 at full augmentation.

service. During training missions at Luke AFB, aircraft sometimes had to wait for 45 minutes or more before takeoff, with engines running at idle settings. This gradually heated the mass of fuel in the Eagle's tanks to the point where it was no longer cold enough to cool the engine control unit.

Modified fuel flow

Given the high temperature of the desert environment at Luke, the unit could not radiate the excess heat away, so tended to overheat. This resulted in engine overspeed problems and turbine failures. The fix in this case, according to the Burrows/Shawler paper presented to the Society of Experimental Test Pilots in 1975, was to cut back the fuel flow rate of the electronic engine control, based on the rate of temperature increase, just prior to full military power. Acceleration time was reduced by this modification to less than 50 per cent of the original time.

Another early problem was that the afterburner light-up envelope was considered too restrictive. Among the more important changes was one involving the five-segment afterburner range (practically mandatory for all fan engines) designed to make a smooth transition between segments, where the

Right: Test equipment is positioned on an F100 at Pratt & Whitney's Government Products Division, West Palm Beach, Florida.

main problems were either too rich a mixture causing a light and then a blow-out, or too lean a mixture preventing a relight at all.

The spray-ring of each segment of the afterburner has a quick-fill capability which controls the light-up of each segment. The amount of fuel in each segment was the same for all altitudes and speeds, resulting in insufficient fuel at sea level and too much at high altitudes. The solution was the provision of a barometric sensor to reduce the amount of fuel fed into the quick-fill area as altitude was increased. The improvement is described as 'the biggest increment' on this problem.

The other major afterburner change

was to reduce fuel flow in full reheat mode, which countered the effects of excess fuel for airflow and afterburner size. By reducing the total fuel flow by a nominal 4,000lb/hr (1,814kg/hr), the blow-out difficulty at full reheat was eliminated. An additional benefit was equal or greater thrust for less fuel flow.

The limited throttle movement came in for some criticism at this stage of the flight testing. Originally, the rpm was to be kept to a minimum at all altitudes, thus allowing 65 per cent rpm at 40,000ft (12,000m). This caused problems in getting the engine out of idle, due to low fuel flow, while requiring a slow throttle movement for acceptable acceleration. The fix was simple enough, involving

increasing the idling rpm in line with increasing altitude.

On the air intakes, which incorporate a variable-geometry ramp to adjust the airflow at varying speeds and heights to the optimum required by the engine, months of wind-tunnel research paid off. The amount of travel provided for the ramp was found to be more than necessary, and the engine inlet compatibility testing was completed in four months. Similar testing with the F-4 Phantom took 22 months, and with the F-101 Voodoo some 30 months.

Early operational and durability problems with the F100 during the late 1970s were largely overcome by modifications plus improvements in materials, maintenance and operating procedures. Production of spare parts was accelerated, and field maintenance teams were increased in size.

Part of the problem lay in the fact that the USAF had underestimated the number of cycles which engines aboard such high-performance types as the F-15 and F-16 would actually undergo. (A cycle is defined as the temperature variation experienced in a mission from engine start to maximum power and afterburner, then back to the lower settings used for landing). In 1977 the service estimated that each engine would undergo 1.15 cycles per flight hour. In practice the rate was 2.2 for the F-15 and 3.1 for the F-16.

At one time, designers had assumed that the most arduous duty which a jet engine had to face was running for long periods at high power levels. By the late

1960s, research had shown that this was simply not the case. Many failures were due to this type of running, but others were created by the heating and cooling resulting from an engine being run up to high power then throttled back.

Technicians dubbed this 'low-cycle fatigue', but had to admit that it was difficult to measure. To aid the design of future engines such as the F100, estimates were made of the average number of thermal cycles to which an engine would be exposed per flying hour. Unfortunately for the F100 programme, these estimates were wrong. In practice, engines were being subjected to far more thermal cycles than the designers had allowed for.

Overworked engines

Paradoxically, the additional stress which the engines were receiving was largely due to the F-15 and F-16 being such good aircraft. Given the high manoeuvrability of their new mounts, pilots were flying in a manner not possible on earlier types, pushing the aircraft to high angles of attack and making full use of the extended performance envelope. In the heat of a dogfight, the throttle setting would be changed much more often than on earlier fighters. All this spelled hard work for the engine.

Critical components such as first-stage turbine blades showed signs of

distress, condemnation rate during repair being 60 per cent instead of the predicted 20 per cent. Maximum gas temperature was reduced to conserve component life, while R&D funding was concentrated on improvements to reliability rather than increasing thrust. Despite these problems, the F-15 had a better engine related safety record by the end of the 1970s than any other USAF fighter at a comparable point in its service career.

Another problem which was to dog the F100 during the first years of its service career was stagnation stalling. The compressor blades in a jet engine are of aerofoil section, and, like the

Above: The split tail of the F-15 allows the stabilators and vertical stabilizers to be kept well clear of the efflux from the engines.

wing of an aircraft, can be stalled if the angle at which the airflow strikes them exceeds a critical value. Powerplant stalls are occasional occurrences in most jet engines, particularly in the early stages of development, and the F100 was to prove excessively vulnerable to stagnation stalling during its first few years of operational service.

Turbofans are prone to a particularly severe type of stall from which recovery is not possible. As the flow of air through

Below: The titanium frames of an F-15 engine bay, with the powerplant about to be installed, as seen by the wide-angle lens.

24

the compressor is disturbed, the engine core looses speed, while the combustor section of the engine continues to pass hot gas to the turbine, causing the latter to overheat. If this condition is not noticed, the turbine may be damaged.

Experience with the F-15 showed that in the event of a mild hard start, the pilot might not notice that a stall had occurred, as the loss of acceleration on the twin-engined aircraft was often not sharp enough to indicate to the pilot that one engine had failed. Unless he checked the temperature gauge, low-pressure turbine entry temperature could reach the point were damage might occur. To avoid this problem, an audible-warning system was devised.

Some stagnation stalls were found to be due to component failures, but most were linked with afterburner problems. The latter usually took the form of 'hard starts' – virtually mini-explosions within the afterburner. In some cases the afterburner failed to light on schedule; in other instances the burner extinguished. In either event, large amounts of unburned fuel were sprayed into the jetpipe, creating a momentary build-up of fuel. When this was ignited by the hot efflux from the engine core, a pressure pulse was created – the aerospace equivalent of a car backfiring.

Deliberate hard start

A reporter from the journal *Aviation Week* gave this account of a deliberately-induced hard start on a test stand: "The force of the auto-ignition was sufficient to rock the heavily sound-insulated concrete test building. A large gout of flame at the afterburner exhaust was seen on the closed-circuit color-television system."

The pressure in the afterburner resulting from a hard start sent a shock wave back up through the fan duct. When this reached the front section of the engine, it could cause the fan to stall, the high-pressure compressor to stall, or, in the worst case, both. It was sometimes possible for a series of stagnation stalls to occur, with each resulting in the afterburner hard start needed to trigger off another.

Stagnation stalls usually took place at altitude and at high Mach numbers, but rarely below 20,000ft (6,100m). Normal recovery method was for the pilot to shut down the engine, and allow it to spool down. Once the tachometer showed that engine rpm had fallen below the 50 per cent mark, the throttle could safely be reopened to the idle position, and the F100 would carry out its automatic relight sequence. Critical factor in restarting the engine after a stagnation stall is the low-pressure turbine-inlet temperature. This must fall to 450deg F (232deg C) before the engine can be restarted.

Several modifications were devised to reduce the frequency of stagnation stalls. The first approach taken was to try to prevent pressure build-ups in the afterburner. A quartz window in the side of the afterburner assembly allowed a flame sensor to monitor the pilot flame of the augmentor. If this went out, the flow of fuel to the outer sections of the burner was stopped.

When the F100 engine-control system was originally designed, P&W engineers allowed for the possibility that ingestion of efflux from missiles might stall the engine. A 'rocket fire' facility was designed into the controls in order to cope with such an eventuality. When missiles were fired, an electronic signal could be sent to the unified fuel control system which supplies fuel to the engine core and to the afterburner. The angle of the variable stator blades in the engine could be altered to avoid a stall,

while the fuel flow to the engine was momentarily reduced, and the afterburner exhaust was increased in area to reduce the magnitude of any pressure pulse in the afterburner.

Tests had shown that the 'rocket fire' facility was not needed, but P&W engineers were able to use it as a means of preventing stagnation stalls. Engine shaft speed, turbine temperature and the angle of the compressor stator blades are monitored on the F100 by a digital electronic engine control unit. This normally serves to 'fine-tune' the engine throughout flight to ensure optimum performance.

By monitoring and comparing HP spool speed and fan exhaust temperature, the engine control unit is able to sense that a stagnation stall is about to take place, and send a dummy 'Rocket Fire' signal to the unified fuel control system to initiate the anti-stall measures described above. At the same time, a second modification to the fuel control system reduces the afterburner setting to zone 1 – little more than a pilot light – in order to help reduce pressure within the jetpipe.

In an attempt to prevent any pulses coming forward through the fan duct from affecting the core, P&W engineers devised a modification known as the 'proximate splitter'. This is a forward extension to the internal casing which

splits the incoming airflow coming from the engine compressor fan, passing some to the core of the engine and diverting the remainder down the fan duct, past the core and into the afterburner. By closing the gap between the front end of this casing and the rear of the fan to just under half an inch (1.3cm), the engine designers reduced the size of the path by which the high-pressure pulses from the burner had been reaching the core.

Engines fitted with the proximate splitter were test-flown in the F-15, but this modification was not embodied in the engines of production Eagles, whose twin engines made the loss of a

Above: The 'nodding' air intakes on either side of the F-15's forward fuselage are necessary to maintain the optimum rate of airflow to the engines, and for operation at high angles of attack.

single engine less hazardous. It was, however, fitted to engines destined for the single-engined F-16.

The improvement in reliability as a result of the modifications to the fuel control system and nozzle was dramatic. Back in 1976, the F-15 fleet experienced a stagnation stall rate of 11–12 per 1,000 flying hours: by the end of 1981 this had dropped to 1.5 per 1,000 hours.

Below: Eagles rendezvous with a KC-10A tanker. Specific fuel consumption of the F100 is 2.17 with full augmentation.

Above: A two-seat F-15 approaches a tanker boom for refuelling. Economical at normal ratings, the F100 consumes 860lb (390kg) of fuel per minute in full afterburner.

Efforts were also undertaken to reduce the smoke output of the F100 as part of a planned component-improvement programme. For example, the combustor was modified to increase the velocity of the airflow in its front end, resulting in improved mixing of air and fuel and leading to more complete combustion and less residual smoke.

Traditional engine-servicing techniques involve replacing critical components at the end of a statistically calculated lifetime. This often results in components being removed and scrapped while still perfectly serviceable, giving good safety margins, but at a high cost to the operator. Engine designers have developed engine parts with greater tolerance to crack damage

so that these may be left in the engine until inspection by non-destructive test (NDT) methods shows that cracks are starting to develop and a replacement is needed. Life-cycle costs may be cut by up to 60 per cent.

The service's Damage Tolerant Design (DTD) programme involved both Pratt & Whitney and General Electric, and focussed much of its attention on the F100. One of the programme's first achievements was a new pattern of F100 fan disc having five times the life of the original component. Key design elements under DTD are high quality control of the raw material, and the avoidance of shapes and configurations such as sharp radii which cause stress concentrations.

Competing against the F220 engine

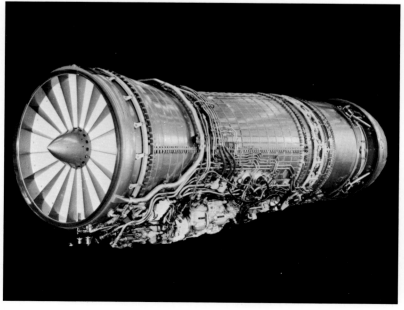

Below left: The General Electric F110 has been test-flown in an F-15, and is one candidate to power future variants of the Eagle.

Top: The P&W PW1120 is based on the core of the F100: smaller and lighter, it is to be used in the Israel Aircraft Industries Lavi fighter.

Above: The other potential replacement for the F100 is P&W's own PW1128, which began test-flying in March 1983.

back in the mid-80s was the General Electric F110. Indeed one of these engines was installed in an F-15 and flight tested at Edwards AFB. The decision to stay with the Pratt & Whitney however, was based upon that company's ability to produce a highly reliable, low maintenance engine.

The Pratt & Whitney F100-PW-220 has been upgraded and developed into the F-100-PW-229.

The engine has been deployed operationally on the F15Es, re-equipping the 21st TFW, Elmendorf AFB, Alaska from September 16, 1991.

Right: Development of the Pratt & Whitney F-100-PW-229 came to a successful fruition when the engine was deployed with the 21st TFW.

F100-PW-229

Avionics

The key to the Eagle's combat capability is its sophisticated suite of avionics: long-range, look-down radar that can detect even low-flying targets at ranges of up to 100 miles (160km); a tactical electronic warfare system to warn of any threat; and displays and controls that present the pilot with almost all the information he needs and allow him to control weapon systems and radar without looking inside the cockpit. The addition of programmable signal processing has further increased the radar capability, and for the ground-attack role still more improvements are planned.

When considering the anatomy of a modern aeroplane the airframe can be likened to the skeleton and flesh, the engines provide muscle and the on-board computers and avionics can be considered as the brain and nervous system. For many years the avionic systems were relatively simple and were purely add-on items to assist in communication and navigation: failure of any one item did not create any great problem, and the aeroplane could fly quite happily under human control and judgement. Then came the great electronic revolution and the emergence of the silicon chip.

The evolution of the integrated circuit – an entire microelectronic system embodied in a chip of semiconductive material, normally crystalline silicon – brought immediate gains to the designers of avionic equipment. The size and weight of 'black boxes' could be drastically reduced; power consumption fell, the associated need for complicated cooling systems; reliability increased; and, perhaps most significantly, the new devices allowed for a considerable expansion of functions.

Computers that had once required large storage space shrank to shoe-box size, offered great reliability and could be programmed for a variety of com-plex tasks. With programming came the new word 'software', denoting an arcane, subjective art that can achieve the apparently impossible.

Generally, software can be considered as the process of telling a computer what to do and how to do it, while making the process understandable to the computer itself. This last consideration is crucial to efficient operation, since a computer totally lacks judgement and intuition and cannot relate in any useful manner to past experience. Consequently, the software must be able to dictate behaviour under all operating conditions including abnormal situations.

High-speed operation

A major advantage of electronic systems is that they operate at the speed of light, allowing a considerable number of operations to take place in a very short space of time. If a pilot operates a switch which affects a circuit under computer control and at that point the computer is working hard it may introduce a delay of perhaps one tenth of a second. This is a valuable breathing space for the computer, yet so short as to be entirely unnoticeable to the pilot.

As well as operational functions, software programs can be used to carry out automatic test and system monitoring, constantly evaluating the health of the associated hardware and providing an indication of malfunction to the pilot. This is an extremely important part of a modern avionic system, usually referred to as built-in test equipment (BITE). It enables failures to be identified and isolated, allowing the pilot to carry on with the operation by selecting alternate systems. One manufacturer of such equipment is on record as saying that a typical system with 100,000 computer instructions may use almost three quarters of this capacity for diagnostic purposes.

As the electronic revolution gained pace so avionic equipment became more complex. By the early 1970s systems were evolving with considerable computational powers, and with digital computers forming an integral part. Reliability had improved and maintenance had been simplified by the use of self-checking systems. More and more critical functions came under the control of electronic systems and the avionic suite now ranked alongside airframe and engine as an essential and fundamental element. Integration of these three areas has reached the point where it is now inconceivable that any part of a modern military aircraft should not be under some form of electronic control of influence.

A modern aeroplane such as the F-15 depends upon electronic systems for communication, navigation, flight management, weapons management, automatic flight control (auto-pilot), systems control and management (eg hydraulics and pressurization), control and management of electronic warfare (EW) systems and the continual monitoring of all aspects of engine and airframe operation and maintenance. Indeed, the pace of change in modern technology and military tactics to meet evolving threats is such that any simple list of electronic devices is necessarily incomplete. Although the aeroplane appears externally unchanged, its black boxes can undergo constant evolution and refinement, and the practised eye can sometimes note the addition of certain lumps, bulges and antennas on the external surface of the airframe.

Above: The McDonnell Douglas head-up display present all essential navigation and attack information to the Eagle pilot.

Below: Nose radars, HUDs and bubble canopies give these F-15s unmatched target detection ability.

Air-to-air gun mode

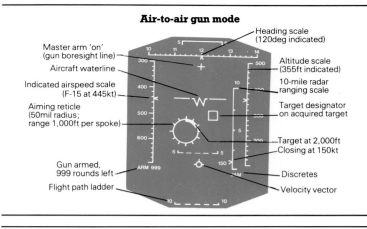

Master arm 'on' (gun boresight line)
Aircraft waterline
Indicated airspeed scale (F-15 at 445kt)
Aiming reticle (50mil radius; range 1,000ft per spoke)
Gun armed, 999 rounds left
Flight path ladder

Heading scale (120deg indicated)
Altitude scale (355ft indicated)
10-mile radar ranging scale
Target designator on acquired target
Target at 2,000ft
Closing at 150kt
Discretes
Velocity vector

Air-to-air medium range missile mode

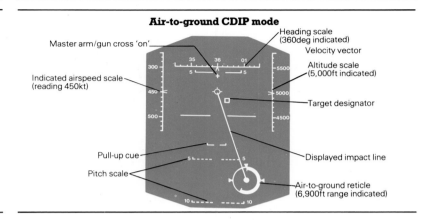

Master arm 'on' (gun boresight line)
Steering circle
Steering dot
Indicated airspeed scale (F-15 at 445kt)
Velocity vector
Missiles armed, 4 MRMs ready
Flight path ladder

Heading scale (120deg indicated)
Target designator
40-mile radar ranging scale
Altitude scale (355ft indicated)
Target closing at 950kt at 22 miles
Target in missile range
60sec to missile impact

Air-to-ground automatic mode

Master arm/gun cross 'on'
Release cue
Indicated airspeed scale (reading 450kt)
Azimuth steering line
Pitch scale

Heading scale (360deg indicated)
Altitude scale (5,000ft indicated)
Air-to-ground reticle (10,000ft range indicated)
Target designator
Time-to-go
Pull-up cue

Air-to-ground CDIP mode

Master arm/gun cross 'on'
Indicated airspeed scale (reading 450kt)
Pull-up cue
Pitch scale

Heading scale (360deg indicated)
Velocity vector
Altitude scale (5,000ft indicated)
Target designator
Displayed impact line
Air-to-ground reticle (6,900ft range indicated)

Above: Typical displays presented by the F-15 HUD showing the symbology used in various air-to-air and air-to-ground attack modes.

This is an important aspect of avionic innovation as it can enhance abilities and performance even though the aircraft is still restricted by its own basic dynamic performance. In other words, the power/weight ratio tends to remain constant, but its electronic eyes and brain can become more far-seeing and powerful so that it can be a more effective fighting machine.

The basic electrical power of the F-15 is provided by engine driven generators manufactured by the Lear Siegler Power Equipment Division. These feature a 40/50kVA generator constant speed drive unit, produced by the Sundstrand Corporation, which ensures that the generator itself is always driven at a constant speed regardless of engine speed or revolutions. This in turn ensures a constant, steady output of closely controlled electrical power in terms of voltage, frequency and phase and does away with the need for additional on-board devices to ensure such

basic integrity of power supplies. Power is then distributed throughout the aircraft via other control systems, circuit breakers, distribution boards and transformer-rectifiers to ensure adequate power, of the correct type, for each sequence of operation. For instance, certain switched selection circuits can accept fairly brutal power sources while other types of instrumentation require highly accurate and sensitive power inputs.

Much of the flight information can be presented to the pilot on a cathode ray tube (CRT) display, which can accept

inputs from radar or electro-optical sensors. This is in line with the current trend for information to be presented in visual terms, often supported by colour. The pilot can assimilate more information more efficiently and ambiguity is reduced. An IBM on-board digital air data computer is used to process information from other sensors such as al-

Below: View through the head-up display following the launch of a medium-range AIM-7 Sparrow. The missile itself is visible in the target designator box.

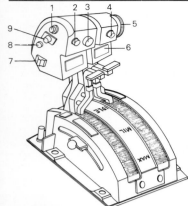

Left: The front panel of the F-15C cockpit has all the most frequently used controls conveniently mounted and a minimum of instruments.

1 Microphone switch
2 IFF interrogate button
3 Target designate control
4 Gunsight reticle stiffen/reject short-range missile
5 Radar antenna elevation control
6 ECM dispenser switch
7 Weapon selection switch (gun, short-range missile or medium-range missile)
8 Spare
9 Speed brake switch

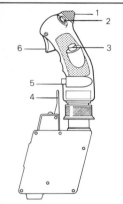

1 Trim button
2 Weapon release button
3 Radar auto acquisition switch
4 Autopilot/nose gear steering release switch
5 SRM/EO weapon seeker head cage/uncage control
6 HUD camera and gun trigger

Above: The throttles (top) and control stick carry all the weapons and radar controls needed in combat, enabling the pilot to keep his eyes attention on the target or HUD.

The F-15 carries such a system in the Northrop ALQ-135, which is part of the aircraft's tactical electronic warfare system (TEWS) and is associated with the active jamming role.

More readily apparent is the Loral ALR-56 radar warning receiver, four external antennas mounted at each wing tip and on top of each fin give them a distinctive, easily recognizable shape. A fifth blade-shaped antenna is mounted under the forward fuselage. Associated equipment includes receivers, power supply, receiver controls and a display of an alpha-numeric type which indicates the degree of lethality and range of the emerging threat. The all-solid-state ALR-56 is based on a digitally controlled dual channel receiver which scans from H-band through to J-band (6-20GHz), while changes in the perceived threat can be accommodated by changes in software.

The APG-63 radar

While it is invidious to suggest that any one part of the avionic system is more important than any other in these days of integrated and interrelated systems, it is hard not to admit that the awesome capability of the Hughes APG-63 radar is really the heart of the F-15 and the

timeters, and this too is presented on the CRT display.

Communication is usually through VHF/UHF links, which essentially form a line-of-sight system. In other words, the range of the equipment is directly proportional to the aircraft's height above the ground. For long-distance flights, therefore, the communications system is supplemented by HF to provide the necessary range. Interference by the enemy is always a problem in communications and several programmes have been undertaken to overcome the problem. Typically, the signal can be spread over a wide spectrum to reduce the chances of detection and require a widely spread jamming signal with consequent dissipation of power and loss of effectiveness.

The aircraft's Litton ASN-109 navigation system is based upon inertial navigation techniques. This is a completely passive, on-board system which does not have to rely on ground-based aids and is largely automatic in operation. It depends upon highly sensitive gyroscopes which are used to accurately align a platform in relation to true North. The pilot tells the computer the start point and can add several desired destinations or waypoints, and as the aircraft moves off, accelerometers on the platform detect rates of movement. All the information is then processed by a digital computer, which comes up with a variety of answers which are displayed to the pilot. This information includes position, time to go to next waypoint and wind conditions.

The high degree of accuracy of platform stabilization can give attitude information for the aircraft's flight instruments providing pitch, roll and heading data at all times. This navigation system is backed up by ground-based navaids such as TACAN, ADF and ILS allowing the F-15 to integrate with any type of traffic pattern. In a cross-country mode these aids can be used to update the inertial system.

Among the most classified equipment is that concerned with aspects of electronic warfare (EW). This is also likely to be the system most often modified, changed and reprogrammed. EW systems can be used to detect radars, notify the pilot of various types of hostile EW and allow a degree of offensive reaction such as jamming enemy signals. Such is the state of modern electronics that many of these devices act completely automatically and can recognize the difference between friendly and hostile emissions. They are capable of reacting to rapid changes in the enemy scenario and can fire off decoys such as chaff or infra-red (IR) flares to confuse the seeker of enemy missiles.

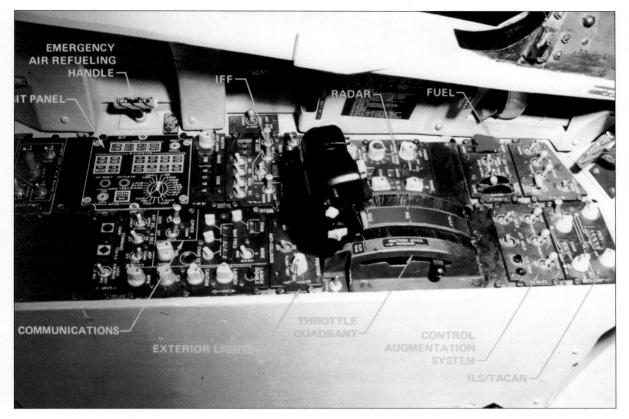

EMERGENCY
AIR REFUELING
HANDLE

IFF

RADAR

FUEL

IT PANEL

COMMUNICATIONS

EXTERIOR LIGHTS

THROTTLE
QUADRANT

CONTROL
AUGMENTATION
SYSTEM

ILS/TACAN

Left: A feature of the left console is the BIT (built-in-test) equipment panel, allowing the pilot to locate faults in the avionics.

Below left: Adjusting the head-up display controls of an F-15D during pre-flight checks at Kadena Air Base, Okinawa.

foundation of its air combat efficiency.

When the F-15 was designed its primary mission of air superiority depended on an advanced fire control radar. The daunting specification numbered among its requirements: use in a single-seat aircraft to track targets at extremely long ranges; close-in and look-down operation that would blind other radars; a clutter-free radar display which would show all target information; the ability to provide tracking and steering data on a head up display (HUD) allowing the pilot to keep his eyes on the target; ease of operation; weapons control and coordination; and selected air-to-ground capabilities. In addition it was expected to reach high standards of reliability and maintainability.

Such a requirement would have been unthinkable a few years earlier and showed just how much influence the new electronics could have on the capabilities of a new aircraft. Even so, to meet the specification certain compromises had to be reached and innovative techniques employed to the full.

Most airborne radars work in the X-band (8–12.5GHz) and choice of frequency is the first area of compromise. The critical factor in an aircraft is the size of the antenna: in a fighter the most convenient position is in the nose, where space is restricted by aerodynamic considerations. It so happens that X band produces a good compromise antenna size. Dropping to S band demands a larger antenna, while going further up, say to K band, offers a small and neater antenna but a signal which is adversely affected by meteorological conditions such as rain – hardly a good choice for a modern fighter aircraft.

Pulse repition frequency

Another fundamental choice is that of the pulse repetition frequency (PRF) which refers to the number of transmitted pulses per second. This is often classed as high or low, with high considered to be energy transmitted at 100,000 or more pulses per second while low is only some 1,000 pulses per second. In general terms, the use of a high PRF in the pulse-Doppler radars common in fighters gives good long-range detection of head-on targets, but a restricted detection of tail-on targets and a tendency to lose track of manoeuvring targets.

In comparison, the low-PRF radars then commonly employed for air combat proved to be good for ground mapping but could not detect targets in a look-down mode. It seemed that the compromise choice of a medium PRF would offer improved performance against manoeuvring targets, but at the expense of long range detection.

Hughes overcame the difficulty by developing a radar that had all three PRF modes. High and medium operate together, while low is used for ground mapping. So emerged the APG-63 multimode pulse-Doppler radar with an all-altitude, all-aspect attack capability and a maximum detection range in excess of 100 miles (160km). It can also guide radar-controlled missiles against all

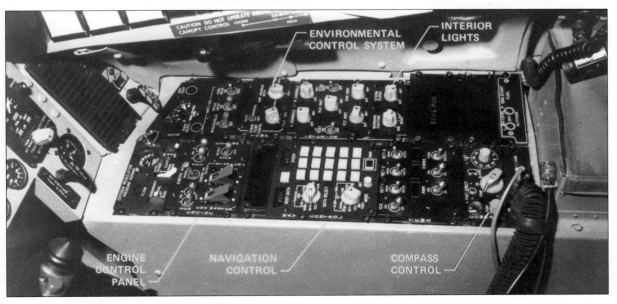

CAUTION DO NOT OPERATE
CANOPY CONTROL

ENVIRONMENTAL
CONTROL SYSTEM

INTERIOR
LIGHTS

ENGINE
CONTROL
PANEL

NAVIGATION
CONTROL

COMPASS
CONTROL

Left: Layout of the navigation, engine and environmental system controls on the F-15 cockpit's right console.

Below: The radar display format used in the Strike Eagle includes a window which can be moved to select enlarged patch maps.

Bottom: Comparative effective ranges of the F-15's APG-63 and the MiG-25's Fox Fire radars, showing the former's clear advantage.

Above: Scott AFB, Illinois, as seen by the camera (right) and as an 8.5ft (2.6m) resolution ground map produced by the F-15 AFCD radar.

types of target. The primary radar controls are mounted on the aircraft's throttle so allowing the pilot to keep his head up during combat. The HUD shows target positioning, steering, range data and weapons release data.

The radar display provides a clear, clutter-free presentation and look-down view of any target aircraft even in the presence of heavy ground clutter. This ability is the result of a combination of both high and low PRF; digital processing of data, and the use of Kalman filtering in the tracking loops. The last process is a computer technique which continually compares the relative errors of the on-board equipment and data from the external sensors – in other words, a form of averaging of available data. In addition, the radar uses a gridded travelling wave tube that permits variation of the radar waveform to suit the prevailing situation.

Another problem with radar is that of false alarms. These are eliminated by the use of a low-sidelobe antenna – which is a good preventative against enemy EW – a guard receiver and frequency rejection of ground clutter and ground moving targets. Air-to-ground modes include target ranging for automated weapon release, a mapping mode for navigation and, thanks to the Doppler element, a velocity update for the INS. The nose-mounted antenna is a planar array that is hydraulically driven and gimballed in three axes.

Combat evaluation

When the radar first came into operation in the F-15 a series of air combat evaluations were carried out by USAF pilots flying the aircraft against seven different types of aircraft modified to simulate leading threats, and the F-15 won by a handsome margin. More significantly, perhaps, it took part in a series of exercises in which an E-3A Sentry AWACS faces attack by a large force. The F-15 proved to be successful in 38 out of 39 intercepts and its radar overcame the effects of jamming techniques.

Further improvement came with the

development of a programmable signal processor (PSP). This is a feature of all new APG-63s and is available as a retrofit item to earlier models. When it appeared, in 1979, this was the only known deployed PSP and was considered to be a key element in expanding the F-15's tactical air interdiction role while enhancing its tactical air superiority capability. The PSP is a high-speed, special-purpose computer which controls the radar modes through its software rather than through a hard wired circuit design; this allows rapid switching of modes for maximum operational flexibility. PSP-modified radar is fitted to the 420th and subsequent Eagles, which are designated F-15C (single-seat) and F-16D (two-seat).

The use of the PSP paved the way for the modification of an ANAPG-63 to provide synthetic aperture radar (SAR). The modification followed on from earlier SAR work in the USAF Forward-looking Advanced Multimode Radar (FLAMR) programme. The PSP in this instance is a fourth-generation model which performs over seven million operations per second.

SAR imagery sharpens mapping details and provides the pilot with an overhead view, as if he were flying directly over the target, when in fact he can be 100 miles (160km) away. Previously, such imagery had to be processed on the ground because suitable equipment was too large to be easily fitted into an aircraft and processing speed was too slow for real time display.

The first flight of the new radar in an F-15 took place in November 1980, and initially a radar mapping resolution of 127ft (39m) was obtained. Within a month it was down to 60ft (18m) – still not good enough for the recognition of small tactical targets. By the 40th test flight, however, resolution was down to the stipulated level of just 10ft (3m).

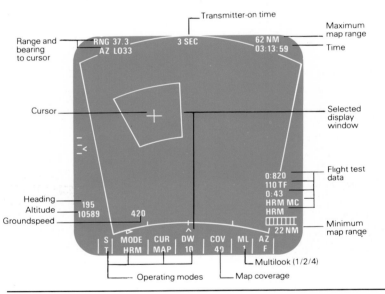

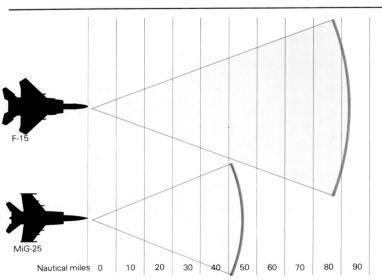

This degree of resolution means that at a distance of 20–30 miles (30–50km) from the target street patterns, power lines and field boundaries are visible. At 10 miles (16km) from an airfield a ⅔-mile (1km) square radar map can be displayed and aircraft as little as 8ft 6in (2.6m) apart can be clearly recognized. Other targets may be seen equally clearly, making target selection easy and unambiguous. The radar maps are updated every few seconds and linked to the navigation systems and weapons modes for ground attack preparations.

By this time, too, MCAIR had developed low-drag fuel pallets to increase the F-15's ferry range. These are two close fitting packs known as fuel and sensor tactical (FAST) packs. In addition to increased fuel tankage, the FAST packs allow a greater range of electronic sensors to be carried, including optical cameras, low light level TV cameras and a laser designator.

Enhanced Eagles

Coincident with this development, the manufacturers suggested that the USAF needed an all-weather fighter capable of performing long-range, air-to-ground interdiction missions while maintaining its air-to-air capabilities, and suggested than an enhanced Eagle would fulfil these requirements.

Two-seater F-15 Eagles had been built and were externally identifiable only by their larger cockpit canopy. This allowed them to function as trainers as well as being combat capable. The manufacturer used one of these two-seaters in a company-funded project known as Strike Eagle to create an all-weather, day/night ground attack aircraft using the new SAR radar integrated with forward looking infra-red (FLIR) system of the Pave Tack pod. The rear cockpit was modified to allow a specialist crew member to handle the radar and FLIR inputs.

The rear cockpit was fitted with four electronic displays and two hand controllers which allow the crew member to focus his attention on the displays while operating systems and controlling

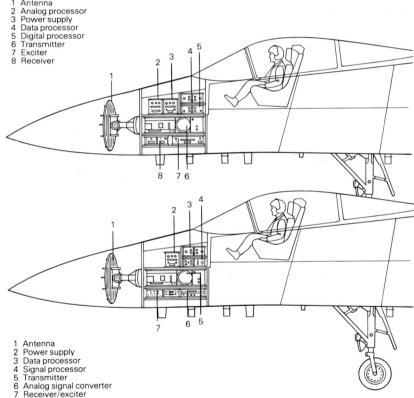

1 Antenna
2 Analog processor
3 Power supply
4 Data processor
5 Digital processor
6 Transmitter
7 Exciter
8 Receiver

1 Antenna
2 Power supply
3 Data processor
4 Signal processor
5 Transmitter
6 Analog signal converter
7 Receiver/exciter

Above: Comparison of the standard APG-63 (top) and the PSP-modified equipment carried by the F-15C.

display content. Two of the displays are used for navigational purposes, one for weapon selection and one to monitor enemy tracking systems.

At the same time, improving electronic systems had created more sensitive electronic countermeasures sensors and the manufacturers proposed an Advanced Wild Weasel F-15. The outcome of these projects is described elsewhere in this book though they deserve mention in the evolution of the aircraft's avionic systems.

In 1981, another advanced avionic feature became linked with the F-15,

when the Integrated Flight/Fire Control (IFFC) and Firefly III programmes were initiated. The IFFC 1 programme, being undertaken by MCAIR under a $14 million contract from the Air Force Flight Dynamics Laboratory, is for the design, development, integration and flight testing of a system which couples the Eagle's fire control and flight control systems to accept dual control inputs and tailor flight control response to the various weapons delivery modes. The Firefly III programme, being conducted by the General Electric Aircraft Equipment Division under a $7 million Air Force Avionics Laboratory contract for the further development of the fire control system.

The IFFC/Firefly III coupling will

Above: Routine maintenance on an F-15A's APG-63 radar. All the equipment is readily accessible, and individual components are easily removable for repair.

allow automatic positioning of the aircraft to attack targets detected by an electro-optical target designation pod. This is expected to shorten engagement times and enable the aircraft to drop its bombs or stores without having to overfly the target. The F-15 in the Firefly programme carries a Martin Marietta-built Atlis II designator pod in the port forward missile well and this is linked to the aircraft's fly-by-wire system via an intermediate additional digital computer.

As had been predicted by the manufacturers at the start of the Strike Eagle project, the USAF is becoming acutely aware of its need to have a long-range all-weather aircraft with a good air-to-ground capability. The current F-4 is nearing retirement and the possibility of an all-new multi-role fighter is unlikely to be translated into reality much before the mid-1990s.

A more practical solution would seem to be the upgrading and modification of an existing aircraft, principally through the use of improved avionics. This is now the subject of a USAF programme which is comparing the relative performances of the F-15 and F-16. Three F-15 advanced fighter demonstrators will be used to evaluate both single pilot and pilot/specialist officer crew complements. The single-seater is an F-15C; one two-seat F-15D is fitted with a Ford Aerospace FLIR and laser tracker/marker pod while the other two-seater is the original Strike Eagle with Pave Tack FLIR and laser pod and synthetic aperture radar. The rear cockpit of the last is still configured with an array of special displays and controls for operation of the sensors, and the aircraft is now known as the Advanced Fighter Capability Demonstrator. Although the Pave Tack pod is expected to be the standard fit, it was considered prudent to gain additional operational experi-

Above: The Strike Eagle's APG-63 was modified to use synthetic aperture radar techniques for high resolution ground mapping.

ence by using the Ford Aerospace system, which has been derived from a similar pod developed for the F/A-18 Hornet.

Another alternative option for the enhanced Eagle is the Martin Marietta LANTIRN (Low Altitude Navigation and Targeting Infra-Red for Night) system currently under development for the USAF's F-16 and A-10 aircraft. The system consists of two pods and cockpit displays: the navigation pod contains a wide-field-of-view FLIR sensor, terrain-following radar and associated electronics, while the targeting pod contains a stabilized wide- and narrow-field-of-view targeting FLIR sensor, automatic trackers, laser designator and ranger, automatic target recognizer, missile boresight correlator and supporting electronics. Interface with the pilot is through a wide-field-of-view, diffractive optics HUD, built by Marconi Avionics in the UK. The system is still under development, and its inclusion on an enhanced Eagle is subject to its successful service entry on the F-16 and A-10.

Possibly the greatest operational

Below: Modifications introduced on the IFFC/Firefly III flight and fire control system, compared with the standard F-15 system.

problem that has to be overcome in the advanced role is that of survivability. This can be achieved only by the use of advanced sensors and more capable avionics, probably with a two-man crew, while Visionics will be needed for all-weather, low-altitude flight using terrain-following radar and a new radar altimeter. Some of this innovation will evolve as part of the constant product improvement programme that companies tend to pursue during the lifetime of their products. In the USAF this is epitomized by the Multi-Stage Improvement Programme (MSIP) which is currently monitoring enhanced radar, improved software and the introduction

of Seek Talk and JTIDS.

Seek Talk is the codename for a long term programme to reduce the vulnerability of UHF radios to enemy jamming by modifying existing radios with the addition of spread spectrum techniques and the use of a null steering antenna.

JTIDS stands for Joint Tactical Information Distribution System, an ambitious programme intended to provide a high-capacity, reliable, jam-resistant, secure, digital information distribution system which will allow for a very high degree of interoperability between various elements of deployed forces and command and control centres. An interesting feature of the JTIDS system is a

relevant navigation characteristic: Implicit in the signalling structure is a highly accurate measurement of message time of arrival which is readily convertible to transmitter/receiver range. This could prove to be a most useful device in future tactical roles and in a supplementary role to INS, whose principal virtue is its independence of ground installations.

Although a great deal of current development is obscured by the needs of security it is possible to visualise an advanced version of the F-15 in the required role. Current systems of the Strike Eagle standard could be integrated with the results of the Firefly III programme and with new HUDs such as the Marconi Avionics wide field-of-view diffractive-optics system. These systems would operate in conjunction with advanced versions of missiles, guns and laser-guided bombs. Trials were due to start in 1983 of the AGM-65D IIR Maverick air-to-ground missile and the General Electric 30mm gun pod.

The next event in the evolving avionic scene will be the introduction of the Global Positioning System (GPS), which

Left: The IFFC/Firefly III F-15B, with Martin Marietta ATLIS II optical sensor and tracker pod on the port wing missile pylon.

Below: Typical manoeuvring attack made possible by the IFFC system (bottom) compared with the conventional pop-up attack profile.

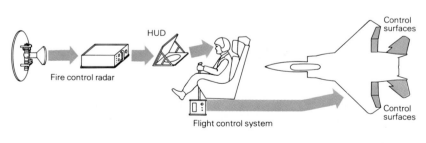

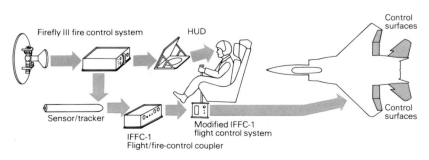

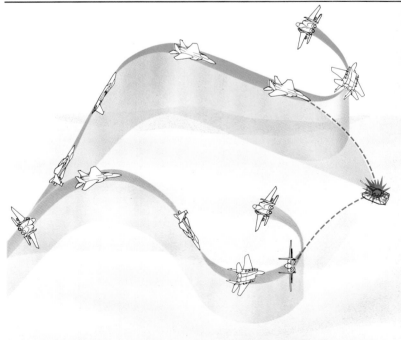

is expected to revolutionize navigation, though it requires satellites which could be vulnerable in war. Currently two F-15s are conducting GPS trials and it is expected that the complete system will be operational by 1987. It will eventually consist of a network of Navstar satellites circling the earth. Thousands of ground receivers mounted in all types of military vehicles and even carried by foot soldiers will translate satellite signals into navigation information that will be accurate to within 10–20m anywhere in the world, day or night and regardless of weather. The information will include altitude, longitude, latitude, velocity to 0.1m/sec and a precise time in nanoseconds.

The Navstar satellites will orbit the earth at an altitude of 12,500 miles (20,000km), each one continuously broadcasting time and position messages. This information will greatly enhance tactical air operations and should prove to be a potent all-weather navigation system. Advanced anti-jamming techniques are built into the system to permit continuing operation under the most stringent of enemy EW operations.

The GPS tactical air configuration will provide continuous signal tracking under all flight conditions and the system will be integrated with other on-board avionics so that the GPS-derived data can be used to refine other flight systems.

Below: Illuminated indicators on the front panel include fire warning (top left), canopy unlocked (top right), air-to-ground mode button and beacon light (centre) landing gear (bottom left) and caution panel (bottom right).

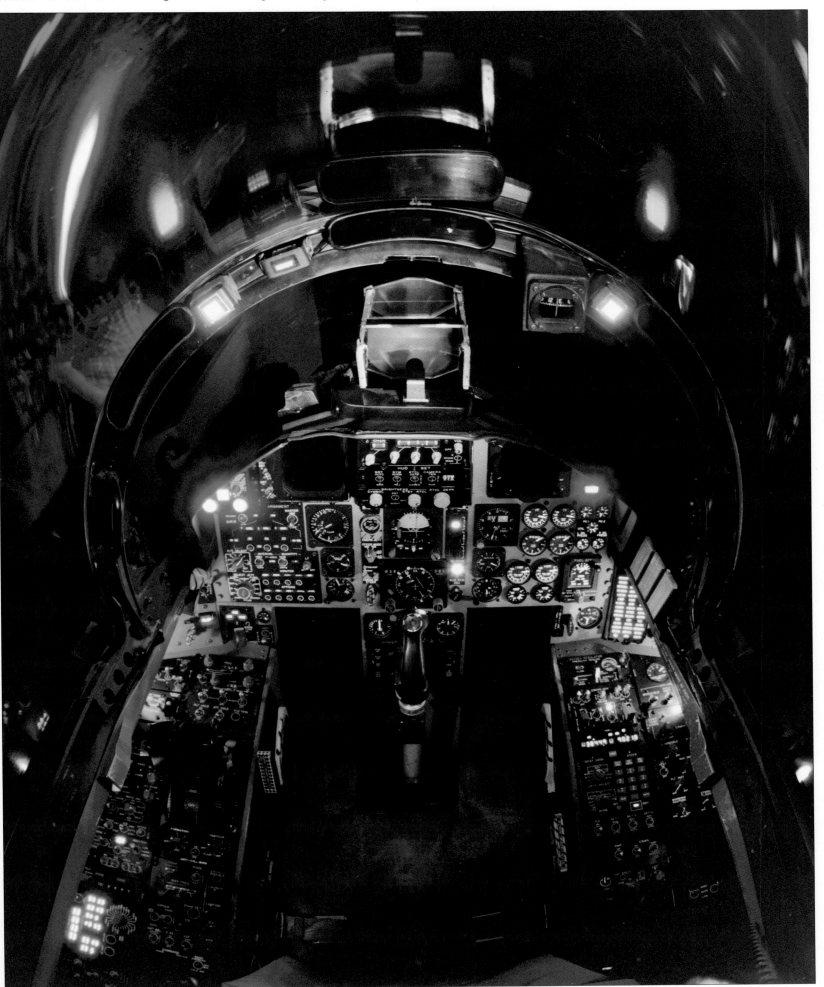

Armament

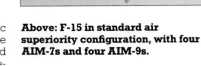

The missiles-only fighters produced for the USAF in the late 1950s and early 1960s were found to be at a severe disadvantage over Vietnam, and the Eagle weapon system was planned from the outset to include a gun. At the same time, improved versions of the medium-range, radar-guided Sparrow and short-range, heat-seeking Sidewinder were provided. With the entry into service of the AIM-120A AMRAAM, and the successful delivery of air-to-ground weapons with a high degree of accuracy, the Eagle is now very much a dual-role fighter.

The basic requirements of the Eagle weapons system, as outlined in the Development Concept fall into three distinct categories: guns, air-to-air missiles and air-to-ground weaponry. In addition, we must also consider the use of external fuel and EW pods to enhance the basic capability of the Eagle.

During the late 1950s and early 1960s air combat with guns was thought by many to have had its day, and the latest American fighters had no guns, relying solely on air-to-air missiles. Air combat during the Vietnam War, however, revealed such an urgent need for gun armament that a new version of the F-4

Phantom, TAC's principal fighter, had to be developed with an internal 20mm gun.

From the beginning, therefore, the FX was planned with an internal gun as an integral part of its weapons system. Initial production models were to rely on the tried and true M61A1 20mm Vulcan cannon, produced by the Aircraft Equipment Division of the General Electric Company. Later models of the Eagle were to have a new 25mm cannon, using a new type of caseless ammunition, which would have the advantage of eliminating cartridge extraction and ejection systems, resulting in a

simpler mechanism. General Electric and Philco-Ford (now Ford Aerospace and Communications) both submitted designs, and after evaluation in December 1971 Philco-Ford was awarded the contract for the new gun, designated the GAU-7.

Although potentially a simpler weapon, the advantages mentioned above were offset by other problems, basically involving the ammunition. With caseless ammunition there is no spent cartridge to be wasted (or collected) after the round is fired, promising potentially enormous savings. However, the US Army had been trying to get caseless ammunition right for the previous 15 years with a continuing lack of success. (The world's first weapon designed with a successful caseless ammunition is the Heckler & Koch G-11 rifle, with ammunition from H&K in collaboration with Dynamit Nobel.)

The GAU-7 ammunition was being developed by the Brunswick Corporation. The propellent half of the 25mm round was covered with a flame retardant covering, which was stripped off mechanically as the rounds entered a

conveyor which took them to the five barrels of the weapon. This stripping mechanism was causing the ammunition conveyor to jam. Another problem involved the development of a moisture-protective material for use with the ammunition, whose final drawback was inconsistent ballistic action. In theory the higher muzzle velocity of 4,000ft/sec (1,200m/sec), compared with the 3,200ft/sec (975m/sec) of the M61, meant that not only did the projectile reach the target sooner than an M61 round, but also that it had a flatter trajectory, producing a more concentrated hit pattern on the target and giving a higher probability of kill. Unfortunately, the whole system proved too unreliable.

By September 1973, with over $100 million spent on development, the problems with the GAU-7 had still to be overcome, and with other new-technology items on the Eagle running reasonably smoothly, the Air Force decided to

Above: F-15 in standard air superiority configuration, with four AIM-7s and four AIM-9s.

Left: Armourers use a conveyor belt to load 20mm ammunition into an F-15 ammunition drum. The links are stripped during the loading process.

Below: Radar-guided Sparrow for BVR engagements and short-range Sidewinders form a potent combination.

Right: Missile launch as seen from the back of a two-seat F-15. At one stage the USAF switched to an all-missile armament, but during the Vietnam war the F-4 had to be revised to accommodate a gun, since many engagements took place at ranges too short for a missile to be launched effectively.

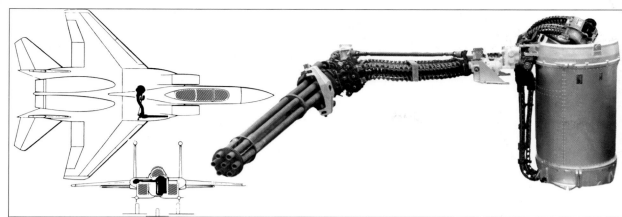

Left: The General Electric M61 Vulcan 20mm gun system, and details of its location in the F-15 airframe. The gun could not be mounted in the nose, since this would interfere with the radar, and the wing root alternative was found to be eminently satisfactory. The ammunition drum, housing 940 rounds, is mounted in the centre fuselage; a linkless feed system transports the rounds to the gun and carries the spent cartridge cases back again. This system has proved extremely reliable in operation.

cut its losses. Although Philco-Ford proposed a year's delay on the programme, it was cancelled and replaced by the M61, of 1954 vintage.

The M61A1 is an exceptionally reliable system. Its rotary action allows a rate of fire of some 6,000 rounds per minute (although only 940 rounds are carried on the Eagle). The use of six barrels minimizes erosion, thus ensuring long life for the weapon. The high rates of fire dictate a special linkless feed for the M61, and while some aircraft systems expel the used cases, others collect them. The ammunition used is in the M50 series, and includes

armour piercing (with or without tracer elements) and incendiary types. The rotary action of the weapon as installed in the F-15 is provided by hydraulic/electrical power.

Locating the gun

The choice of location for the internal gun was not easy. Ideally it should have been placed as close to the fighter's cg as possible in order to reduce aiming errors when the weapon was fired, but a nose mounting was ruled out because the vibration of the weapon firing would upset the microcircuitry of the APG-63 radar. A second location, further aft on

the fuselage centreline, was rejected because of possible gun-gas ingestion problems. The final solution, approached with some trepidation was to mount the gun in the starboard wing root, where there was plenty of room for the weapon and its ammunition drum and feed system. Tests later showed that the comparitively large separation of the gun from the fore-and-aft axis of the aircraft produced no aiming or recoil problems. In addition, the gun alignment could more easily be varied to give maximum tracking time on target, a facility initially demonstrated in simulation, and later proved in practice.

For the future, there is a new General Electric 25mm cannon, the GAU-12/U, under development for the AV-8B Harrier II/Harrier GR.5. Present plans do not call for its use beyond the AV-8B, but the possibility exists that it might be adapted for installation in the F-15E. Another gun option for the F-15E is the carriage of three General Electric GE 430 GEPOD-30 30mm gun pods on fuselage and underwing pylons. The GEPOD-30 is a lightweight four-barrel version of the GAU-8/A gun used in the A-10A Thunderbolt II, and fires the standard GAU-8/A ammunition. The main role of the GEPOD-30 would be

Right: The rocket motor fires as an AIM-7 Sparrow is launched from the first F-15C Eagle. The aircraft has an instrumented nose probe.

air-to-ground, but there is nothing to preclude its use in the air-to-air mode if the tactical situation allows.

Although the gun has been restored, the missile remains the Eagle's primary air-to-air weapon. The new fighter's missile armament was originally to consist of the AIM-7F Sparrow for beyond visual range (BVR) engagements, and a new short-range IR-homing missile, designated AIM-82. The latter missile was cancelled well before it reached the hardware stage, and the AIM-9L Sidewinder took over. For the future, the AIM-120 AMRAAM will replace Sparrow, with the projected European ASRAAM a possible successor to Sidewinder.

AIM-7 Sparrow

The AIM-7 Sparrow originated as Sperry Gyroscope's Project Hot Shot in 1946, and by 1955 it was in service as the beam-riding AAM-N-2 Sparrow I with the US Navy. The active-radar Sparrow II was cancelled in 1957 and the AIM-7F comes from the third generation of Sparrow, the Raytheon AAM-N-6 Sparrow III, which became the AIM-7C when the US services changed nomenclature in 1962. The AIM-7C introduced semi-active radar homing with continuous wave (CW) guidance and was in service by 1958. The AIM-7E, later versions of which armed early Eagles, featured a continuous-rod warhead, consisting of a 66lb (30kg) explosive charge enclosed in a tight drum made from a continuous rod of stainless steel. On detonation, this rod shatters into some 2,500 lethal fragments. The more manoeuvrable AIM-7E2 was developed to reduce the missile's minimum range as a result of experience in Vietnam, when the demand for visual identification of targets inhibited its use, and this version armed the initial batches of F-15 Eagles.

The AIM-7F, designated missile for the Eagle, brought the Sparrow into the solid-state electronic age. Reducing the size and weight of the guidance package, still CW, allowed a more powerful motor, the Hercules Mk 58, to be used, resulting in an increased range of 62 miles (100km), and enabling a larger 88lb (40kg) warhead to be carried. Introduced in 1977, the AIM-7F is claimed to be able to lock-on to a target against clutter up to 10dB.

The most recent version of the Sparrow is the AIM-7M, which has an inverse monopulse seeker, a digital signal processor, a new autopilot and a new fuze. It offers greatly improved results when

Above: The Sparrow accelerates towards its maximum speed of approximately Mach 4. Range of the AIM-7F is 62 miles (100km).

Left: Armourers in protective clothing and respirators install a Sparrow on its fuselage mounting.

Below: In its air superiority configuration, the F-15 can have its fuel, oil and liquid oxygen replenished, and Sparrows and ammunition reloaded, in a turnaround time of only 12 minutes.

fired in adverse weather conditions, as well as in an ECCM environment. Indeed Sparrow production has been concentrated on this variant since FY 1981 budget funding.

AIM-9 Sidewinder

The Sidewinder is the original simple, low-cost air-to-air missile, capable of being carried by practically any combat aircraft. Since its development in 1949 by a team of scientists at the Naval Ordnance Test Station (now the Naval Weapons Centre) at China Lake, it has been produced in seven major variants, with at least one further derivative under consideration. US production of the initial AIM-9B variant reached a total of 80,900 missiles, while a European consortium produced about 15,000. The usefulness of the missile is exemplified by the fact that during the Falklands conflict of 1982 the RAF modified their Nimrod maritime patrol and ASW aircraft to carry four Sidewinders in a matter of weeks in order to give them a measure of self-defence capability.

Sidewinder has an infra-red seeker which homes onto heat emissions from the jet efflux of the target. A variety of guidance heads, and six major forward fin configurations have been used, and many of the later versions combined older airframes with new seekers and fins.

The AIM-9L, representing the third generation of the missile, features a new double-delta forward fin configuration of larger splan than previous missiles and a new seeker head. Part of the DSQ-29 guidance and control system, the new head uses AM-FM conical scan, with a fixed-reticle, tilted-mirror system, and is cooled by the inclusion of Argon gas. The new seeker offers greater sensitivity and improved tracking stability, while lethality is increased by the use of the DSU-15B active optical laser fuze, allied to the WDU-17B annular blast fragmentation warhead.

Once the pilot has energized the missile homing head, he listens for the 'growl' in his headset that signifies it has acquired a target. Should he be set up for a perfect shot straight up the tailpipe of a target, the growl intensifies to a strident tone, and the pilot launches the missile. In this respect Sidewinders (of whatever mark) are simple to use in air-to-air combat.

AIM-120 AMRAAM

Despite all the advances made over the years with the Sparrow AAM, it retains one basic drawback – its semi-active radar guidance system requires the target to be continuously illuminated throughout an engagement, so that the pilot can only deal with one target at a time. Consequently, the prime require-

ment during development of the Advanced Medium Range Air-to-Air Missile (AMRAAM) was that it should be a launch-and-leave missile, allowing several targets to be engaged simultaneously beyond visual range without monopolizing the fire-control radar and leaving the pilot blind to other threats. AMRAAM was also required to be

usable in all weathers, fired from or at all aspects and with a look-down/shoot-down capability. With the proliferation of Sparrow throughout the US Air Force, Navy and Marine Corps, it had to be able to integrate with existing fighters and fire control radar, and physically fit where a Sparrow was previously located. The final requirement was con-

siderably higher reliability than that of Sparrow.

This was a demanding specification, and in 1976 a joint USAF/US Navy project office was set up at Elgin AFB to organize the development of the new missile. By February 1979 the conceptual studies had been narrowed down to submissions from Hughes Aircraft and Raytheon, who were awarded contracts for ten prototype missiles each to be fired from F-14, F-15 and F-16 aircraft in a competitive evaluation.

After only three firings of each contender the trials were halted, with the Hughes entry a clear winner, and in December 1981 the company received a 50-month full scale development contract valued at some $421 million. The resulting missile, designated AIM-120 but still known as AMRAAM, is some two-thirds of the weight of the Sparrow, and similar in configuration, though the main fins are somewhat smaller in size. It has a speed of Mach 4, and an active X-band radar terminal seeker using a high-power solid-state transmitter with a low-sidelobe, wide-gimbal antenna, and a built-in radio-frequency processor. The major operations inside AMRAAM, including navigation, autopilot, radar, datalink, fuzing, sequencing and self-test functions are all handled in the missile by a single 30MKz microprocessor.

The last two of the Hughes test firings were from an F-15, following one from an F-16, and the second launch, on November 23, 1981, scored a direct hit on a QF-102 target drone. The F-15 was flying at 6,000ft (1,830m) at Mach 0.75, and the missile was launched in a look-down/shoot-down mode at the QF-102 flying at 1,000ft (300m) at Mach 0.7. The missile was cued by the APG-63 radar and launched by the aircraft's stores

Top left: A fully armed F-15 stands ready for takeoff, with safety tags on missiles and airframe and cockpit canopy open.

Centre left: An armourer removes the cover from the seeker head of a Sidewinder as the aircraft is prepared for a sortie.

Left: Mock-up of the Hughes AIM-120 AMRAAM is assembled on the fuselage station of an F-15.

Below left: First captive flight test of a Hughes AMRAAM Instrumented Measurement Vehicle, used to test the new missile's aerodynamic compatibility with the F-15.

Below: During the firing trials that resulted in its selection, a prototype of the Hughes AMRAAM is launched from an F-15B.

control system. It demonstrated inertial midcourse guidance and then acquired its target against ground clutter.

The final prototype firing came late in 1982 and demonstrated the missile's ability to intercept a low-flying aircraft using self-screening ECM. It was launched from an F-15 flying at 16,000ft (4,900m) some 10.8nm from a QF-102 drone flying towards the launch aircraft at 400ft (120m). It was launched with command-inertial guidance and then switched to active radar in order to acquire and intercept the target, which passed within lethal range of the AMRAAM's warhead.

Further development firings are scheduled to take place in 1984 from the F-16 (on which the AIM-120 will enter service in 1986) followed by the F-18, F-15 and F-14. The USAF requirement

for AIM-120 is upwards of 20,000 missiles, and already further improvements are planned for the missile's 25-year projected life. The AIM-120B will feature passive terminal homing, and will be introduced from 1990, while the AIM-120C will have overall improvements in range, speed and manoeuvreability, and will be available from the mid-1990s. All variants of the AIM-120 will be able to be launched from either rails or ejection units.

AMRAAM is the subject of an agreement signed between the United States and France, Germany and the UK, which will see the missile become the NATO-standard BVR weapon. In turn, the European partners will develop the Advanced Short Range Air-to-Air Missile (ASRAAM) to replace Sidewinder in both US and European service. Al-

though agreements over ASRAAM are in existence, the missile is unlikely to be available before the late 1980s.

Air-to-ground Weapons
While primarily an air superiority fighter, the Eagle possesses a substantial air-to-ground secondary capability, as the accompanying table illustrates. Work is also under way on a dedicated air-to-ground variant, designated F-15E, which will be a two-seat strike fighter. The basic stores capability of the F-15A/C has five weapons pylons, in addition to the four Sparrow missiles, capable of carrying up to 16,000lb (7,257kg) of bombs, rockets, ECM equipment or external fuel tanks.

The Strike Eagle demonstrator, on which the F-15E will be based, can carry up to 24,000lb (10,886kg) of stores. Among the stores compatible with the Eagle in this role are the AGM-88A High-speed Anti-Radiation Missile (HARM), AGM-65 Maverick TV-guided missile, AGM-84A Harpoon anti-ship missile, Mk 20 Rockeye bombs on MER-200 multiple ejection racks, Matra Durandal runway denial weapons, 500lb (227kg) Mk 82 bombs in Slick (low-drag) and Snakeye (retarded) configuration, 2,000lb (907kg) Mk 84 bombs in Slick, Laser, Electro-optical and Infra-red homing versions, ALQ-131 ECM pods and 600US gal (500Imp gal/2,280 litre) drop tanks.

1 ECM antenna
2 ALQ-119(V) jammer pod
3 600US gal (500Imp gal/2,273 litre) fuel tank
4 MER (multiple ejector rack) carrying three Mk 82 500lb (227kg) slick (low-drag) general-purpose bombs (one with a stand-off contact fuze) plus one AIM-9J and one AIM-9L Sidewinder AAMs
5 FAST (Fuel and sensor tactical) pack conformal fuel tank
6 MK 20 Rockeye cluster bomb
7 Tactical nuclear bomb
8 MK 82 Snakeye high-drag bomb
9 M61 cannon with 940 rounds of 20mm ammunition
10 GBU-10E/B (Mk 84 2,000lb) Paveway II laser-guided bomb
11 AVQ-26 Pave Tack sensor pod

12 GBU-12D/B (Mk 82 500lb) Paveway II laser-guided bomb
13 CBU-52B/B cluster bomb dispenser
14 AIM-7F/M Sparrow AAM
15 AGM-84A Harpoon anti-ship missile
16 SUU-20 practice bomb dispenser
17 MK 84 2,000lb (907kg) general-purpose bomb
18 GBU-15(V)-4-B modular guided glide bomb
19 AGM-88A Harm anti-radar missile
20 AGM-65D IIR (imaging infra-red) Maverick air-to-surface missile
21 Two AGM-65A (TV) or AGM-65C (laser) Mavericks
22 General Electric GPU-5/A gun pod housing 30mm GAU-13/A gun, ammunition and drive system
23 AIM-120 AMRAAM (advanced medium-range air-to-air missile)

The MER-200 multiple ejection bomb rack is designated BRU-26A/A, and allows the Eagle a high degree of flexibility in its weapons carriage. It is in production to equip all Eagles assigned to the US Rapid Deployment Joint Task Force, as well as for the Japanese Air Self Defense Force F-15Js. Its main advantage is that it allows supersonic carriage and release of up to six weapons, and can jettison them in any loading configuration. It is strong enough to allow the pilot to pull 7.3g during combat manoeuvres, and has been flight tested to Mach 1.4, although MCAIR, the manufacturers, claim carriage, jettison and release of stores up to Mach 2. Production of this low drag multiple ejection rack was running at 40 units per month by mid-1983.

In addition, the Eagle can enhance its range by the use of the FAST (Fuel And Sensor Tactical) packs mounted on the fuselage side, and these retain the ability to mount Sparrow or AMRAAM missiles on the lower corners, or carry some 4,400lb (1,996kg) of air-to-ground stores. Each of the FAST packs can carry 849US gal (707Imp gal/3,228 litres) of fuel. Alternatively, or in combination with fuel, they can house cameras and IR sensors for reconnaissance; low-light television (LLTV), forward looking IR (FLIR) and laser designators for the strike role; or Wild Weasel equipment for defence suppression.

FAST packs were first flown on an F-15B on July 27, 1974. MCAIR claim that carriage of the conformal tanks only slightly reduces the subsonic profile drag relative to the clean aircraft and represent only a fraction of the super-sonic drag of the three standard drop tanks, which between them carry 20 per cent more fuel than the FAST packs. Installation or removal is possible in 15 minutes, despite their complex shape Pand size. With FAST packs and the three drop tanks, the F-15C has demonstrated an unrefuelled ferry range of 2,660nm (4,903km). In this configuration, the F-15C has a maximum gross weight of 68,000lb (30,840kg), while the F-15D model is only some 800lb (363kg) heavier with the same internal fuel.

Weapons configurations
In the basic air-to-air role, the F-15A/C Eagle carries four AIM-7F Sparrows, four AIM-9L Sidewinders and the internal 20mm gun. In addition, it usually carries a 600US gal drop tank on the centre-line pylon, and the inner wing pylon can also carry a similar drop tank without sacrificing the Sidewinder capability.

The basic attack configuration of the non-dedicated Eagle retains the basic air-to-air configuration, possibly including the centre-line tank, and adds extra stores. Any range enhancements required can be provided by air-to-air refuelling. The prime requirement was for the attack mission not to detract from the basic air-to-air combat mission, which is certainly the case.

Below: The second F-15B in early 1976, before its development into the Strike Eagle. Even at this stage MCAIR were keen to demonstrate the F-15's ground-attack capability, and 71-0291 is seen here armed with 18 500lb (227kg) Mk 82 slicks.

Left: The impressive array of stores that have been launched by the Eagle or are designed to be compatible with the aircraft's delivery system.

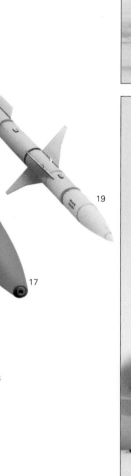

23

22

19

18

17

16

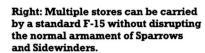
5

Right: Multiple stores can be carried by a standard F-15 without disrupting the normal armament of Sparrows and Sidewinders.

Deployment, Training and Combat

The combat effectiveness of a weapon system is dependent upon its ability to establish and maintain a dominant edge over its adversaries. In the highly demanding arena of air superiority, the ability to deploy superior, reliable technology, together with highly trained, motivated, professional crews, is crucial to the pursuit of that goal. The F-15 and the Squadrons who service and fly it, represent the embodiment of fighting excellence. Indeed, three of the four nations that operate the system have proved this in the ultimate testing scenario – air combat.

Above: An unarmed 36th TFW F-15A on deployment over Norway. The 36th was the first USAFE unit to be equipped with the Eagle.

In due recognition of an outstanding combat record achieved over the skies of Vietnam, it was decided that the 555 Tactical Fighter Squadron (TFS), 'Triple Nickel' would be the first unit to receive the F-15. Its first aircraft, TF-15A, 73-0108, christened 'TAC 1', was accepted during a ceremony presided over by President Gerald Ford on November 14, 1974. Part of the 58 Tactical Fighter Training Wing (TFTW), based at Luke AFB, Arizona, the 555th – now designated Tactical Fighter Training Squadron (TFTS) was manned by Instructor Pilots (IPs), whose task it was to train pilots for front-line Eagle squadrons. In the early days however, this task wasn't as straightforward as it might seem. Engine-related problems such as compressor stalls and stagnations, together with difficulties associated with the fire control system and the APG-63 radar, all conspired

against the training programme to produce low aircraft serviceability levels, which resulted in a pilot shortage. This situation was slightly improved by the decision in November 1975, to eliminate the air-to-ground portion of the training syllabus, but the overall situation was further compounded by the requirement of the 58 TFTW to bail six of its aircraft to the 57th Fighter Weapons Wing (FWW) at Nellis, in order to conduct the Air Intercept Missile Evaluation (AIMVAL) and the Air Combat Evaluation (ACEVAL) portions of the flight test programme.

Despite these trials and tribulations, training of the 1st TFW personnel proceeded at a steady pace, and the first F-15A, 74-0083 'Peninsular Patriot' was delivered to the 27 TFS, at Langley AFB on January 9, 1976. Aircraft deliveries continued and by the end of the year the

other two component squadrons of the wing – the 71st TFS and the 94th TFS had received their allocation of Eagles and were adjudged to be mission-ready.

Two Eagles, 75-0049 and '50 arrived at Bitburg Air Base, Germany on January 5, 1977, these aircraft however, were to be used initially to train maintenance personnel. In a radical departure from previous re-equipment programmes, and under the title Operation Ready Eagle, pilots and other aircraft comprising the 525 TFS, 36 TFW underwent training to operational ready status, under the watchful eyes of the 94th TFS, at Langley AFB. As training continued, two more F-15s joined their stable-mates at Bitburg in March, then on April 27 some 20 F-15As and 3 TF-15As left Langley on a mass transatlantic migration to Bitburg, all participants safely reaching their destination during the afternoon of the same

day. Operation Ready Eagle had proved itself, significantly reducing the 'downtime' usually associated with major unit re-equipment programmes, indeed, within 12 hours of arrival some of the newcomers were 'pulling' Zulu alert duty.

Back at Langley the training process continued and by October both the 53rd and 22nd TFSs were back at Bitburg with their new mounts.

Between June 1977 and July 1978 the 49 TFW, based at Holloman AFB converted its three squadrons from F-4Ds to F-15s under a programme called 'Ready Team'. Under this new initiative some of the wings F-4s remained combat-ready while pilots from the 7th and later 8th and 9th TFS were trained at Luke AFB, by the 58 TFFW (which by now had been re-designated the 58 Tactical Training Wing (TTW) and had itself been

expanded beyond the 555th, to include the 461st and 550 TFTSs). With the 49 TFW becoming the third front-line wing to re-equip with the F-15, Ready Eagle 2, the re-equipment of the 32nd TFS based at Soesterberg in the Netherlands began in May 1978. As this was the only USAF asset at the base, it became necessary to slightly adjust the procedure, in order to honour the unit's defensive commitments to the theatre. Accordingly, exercise Coronet Sandpiper saw the arrival at Soesterberg of 18 F-15s from the 1st TFW, on September 13, 1978. They remained in the Netherlands for approximately 3 months, being manned by 71st TFS pilots for the first 45 days and members of the 94th TFS for the rest of the deployment. Two 32nd TFS, F-15s were supplied during this time for maintenance training, while most of the units aircraft had arrived by December 1978, enabling the Langley aircraft to return home. By mid January 1979 the 32 TFS was fully equipped and operational with 18 A-models and 2 two-seaters.

During the first six months of 1979 the 58th and 59th TFSs of the 33rd TFW at Eglin AFB, Florida traded its F-4s for Eagles and in July of that same year, the 60th TFS was re-activated as part of the Eglin wing and also received F-15s.

Pacific Air Forces (PACAF) were next to benefit from 'Eagleisation' when between July 1979 and April 1980 component squadrons of the 18th TFW – the 12th, 44th and 67th TFSs re-equipped with F-15 Cs and Ds via Ready Eagle 3 made possible by courtesy of the 33 TFW, at Eglin AFB.

The emergence of the F-15 C and D models resulted in the appearance of yet another codename – Ready Switch.

This process took place from between May and August 1980 when Soesterberg traded-up and replaced its As and Bs

with new-build Cs and Ds; a process then repeated next by Bitburg and then Langley and Eglin. The 58 TTW at Lake split away from F-15 training and instead devoted its energies to F-4 crew training. In its place the 405th TTW was re-activated, into which was assigned the 555th, 550th, 461st and 426th TFTSs. The 550th and 461st TFTS currently fly F-15Es, while the other two squadrons are equipped with A, B and D models.

On August 10, 1981 the 48th Fighter Interceptor Squadron (FIS) at Langley AFB received this first F-15A for maintenance and familiarisation training. By the middle of 1982, Langley F-106s had been replaced by F-15s, marking yet another milestone.

Four other Air Defence Tactical Air Command (ADTAC) units flying the F-106 were re-equipped with F-15s shortly thereafter. They were the 5th, 49th, 87th and 318 FISs based at Minot, Griffiss, K. I. Sawyer and the McChord AFBs respectively. Subsequently the 5th FIS has transferred its F-15s to the 101st FIS, Otis Air National Guard Base (ANGB), and the 318 FIS transferred its Eagles to the 123rd, Portland, Oregan ANG. In addition to the 199th FIS operates F15s from Hickham AFB, Hawaii, while tactical air defence is supplied by Air National Guardsmen flying F-15As and Bs with the 122nd TFS at New Orleans Naval Air Station (NAS) and the 128th TFS at Dobbins AFB.

Above: A trio of F-15As from the famous 555th 'Triple Nickel' TFTS, 58th TFTW, based at Luke AFB, Arizona.

The three squadrons making up ADTACs 325 TTW, namely the 1st, 2nd and 95th Fighter Interceptor Training Squadrons (FITSs) at the Air Defence Weapons Centre at Tyndall re-equipped from F-106s, to F-15s and the 57th FIS at Keflavik AB, Iceland traded its F-4Es for C and D model Eagles.

In 1982 the 43rd TFS based at Elmendorf AFB became equipped with 27 F-15As and Bs, and was followed by the 54 FTS – both squadrons have subsequently up-graded to Cs and Ds.

Below left: Ready to get airborne on an interception in less than five minutes, a fully-armed F-15A waits in its hardened shelter.

Below: A pair of F-15As from the 1st Tactical Fighter Wing, the first operational Eagle unit, based at TAC's Langley HQ.

Below: A 1st TFW F-15A banks over one of the world's most expensive flight lines at Langley AFB, Virginia, also home of the 48th FIS.

Bottom: Three F-15As and a two-seat F-15B of the 57th Fighter Weapons Wing, based at Nellis AFB, Arizona, in formation.

Flight Training

For the aspiring F-15 pilot, fresh from pilot training, with no previous experience as a tactical fighter pilot, there are basically two courses that must be successfully completed before the heights can be reached. For entry into the first course known as Lead-In Training (LIT), the student will possess one of the following qualifications.

UPT	Assigned directly from Undergraduate Pilot Training
T-38A (C)	Last rated assignment was a T-38 IP and who was current in the T-38 within 90 days of the first flight of the course
T-38A (NC)	Last rated assignment was a T-38 IP, but was not current within 90 days of the first flight of the course
Other Input	All others

The 43 training days making up the LIT course consists of five ground training and 38 flying training days. On successful graduation from this, the pilot moves to Luke AFB, Arizona. The Operational Training Course (OTC) for F-15 pilots is conducted by the 405th TTW. The OTC syllabus is a modula concept which builds basic fighter skills into each graduate so that on successful completion of the course, the pilot is integrated into the tactical air forces and prepared to fly as wingman in the unit he joins. The course itself spans 195 academic hours over 84 training days, 18 ground training, the rest in the air. In all 52.9 hours of flying is achieved in 41 sorties, with an additional 27 hours logged during 18 sessions in the F-15 simulators. On completion of ground school, the first flight the 'rookie' Eagle Driver logs, covers local orientation and basic aircraft handling. The second entry in his log is proficiency demonstration, after which the student moves to three formation flying blocks – during the second of which the first solo flight is completed. Formation instruction is followed by two instrument training sorties and then the fighter practice begins. Next comes five offensive basic fighter manoeuvres (OBFM) each of 1.1 hours duration, the fifth is a check flight. These are one-vs-one (1v1) with the student trying to gain the advantage over his instructor who is flying as an enemy defender. Back on the ground the instructor thoroughly critiques each flight, using film from the student's gun camera where necessary. These manoeuvres are also practiced in the simulator. Four defensive basic fighter manoeuvres (DBFM) sees the student practicing counters against an enemy using guns and missiles. Next five neutral basic fighter manoeuvres are flown. With both aircraft beginning at about the same altitude, the student moves to radar intercepts beyond visual range (BVR).

The following six intercept sorties are carried out both by day and at night, and involve the student being vectored against high, medium and low targets using Ground Control Intercept (GCI) methods, until the targets are acquired by the F-15s radar at maximum range. These sorties are of longer duration – between 1.4 to 1.6 hours. The next eight one-hour sorties include 1v2 and 2v1 offensive and defensive engagements with visual, radar and GCIs included. After this, four dissimilar air combat sorties are flown, in engagements where the advisories are flying an aircraft other than an F-15. These are fought 1v1, 1v2 and 2v2 using GCI support. In another sortie Combat Air Patrol (CAP) tactics together with their own radar capabilities are used to locate and destroy an enemy force preferably BVR. The final two sorties of the course consist of air-to-air gunnery qualification. On successful completion of the OTC, pilot then reports to his operational squadron, where he will begin the process of achieving mission-ready status.

Eagles for export

The Israeli Defence Force (IDFAF) purchased their first batch of F-15s on December 10, 1976 under a programme called "Peace Fox", and their first custom built aircraft flew on October 12, 1977. Since then other orders have followed, and some F-15s are based at Tel Nof AFB. It is believed that the first squadron to be equipped with the type was 133 Squadron.

The Japanese Air Self-Defence Force (JASDF) considered no less than 13 potential candidates before opting for the F-15. The first F-15 J, aircraft number 544, serial 79-0280 undertook its maiden

Above: Israel was the first export customer for the Eagle, receiving its first three in December 1976. Here four IDF-AF F-15s fly over the historic Masada fortress.

flight on June 4, 1980 and was handed over to JASDF officials at St. Louis on July 15.

Under the watchful eyes of Col. Otsubo, Lt. Col. Hoso and Major Tanake flew F-15Js '280 and '281 extensively, evaluating the aircraft's performance and fire control systems at Edwards and Whiteman AFBs, before Lt. Col. Tom Browning and Maj. Darryl Smith delivered the Eagles in March 1981 to the Air Proving Wing at Gifu AB, Japan, for further test flights. In addition to other F-15Js, and a number of two seat F-15DJs built by MCAIR, Mitsubishi's factory at Komaki also build the type under license. In all six JASDF units are equipped with the F-15 (see table below).

Squadron	Base
202	Nyutabaru
203	Chitose
204	Hyakuri
207	Chitose
305	Hyakuri
301	Nyutabaru

In January 1981 the Royal Saudi Air Force (RSAF) purchased a small number of F-15s and operated them from Luke AFB, as pilot trainers in full USAF insignia. Their first Eagle flew on June 2, 1981, and they began reaching Saudi Arabia the following year, at the end of

Above: A Royal Saudi Air Force F-15C armed with Sidewinders, flies in formation with one of the Lightning interceptors it replaced with No. 13 Fighter Squadron.

which a total of 47 aircraft had been delivered under the Peace Sun programme. Number 13 Squadron from Dharan were the first to receive the Eagle, and were followed by 6 Squadron, at Khamis Mushayt and 5 Squadron, at Taif. Under the 'Camp David' agreement, where it is believed that 60 RSAF F-15s may be based in theatre at any one time, this figure however was increased by a further 24 aircraft at the beginning of Operation Desert Shield. In addition to other F-15s, it is believed that 12 USAFE F-15s from Soesterberg and 12 from Bitburg formed 12 Squadron RSAF also at the base at Dharan.

Into combat

Perhaps not surprisingly it was the IDFAF that drew first blood with the Eagle. On June 27, 1979, a mixed force; consisting of F-15s and Kfirs were providing a CAP for other IDFAF aircraft striking terrorist bases near Sidon, southern Lebanon. An Israeli Hawkeye AEW aircraft detected a number of Syrian MiG-21s rising up to intercept the strike aircraft and directed the CAP against them. In the one sided battle that ensued five MiG's were downed and all the IDFAF aircraft returned home safely. Despite that battle having been fought thirteen years ago, still no further details are available. Two years later, on March 13, 1981, an IDFAF F-15 shot down a

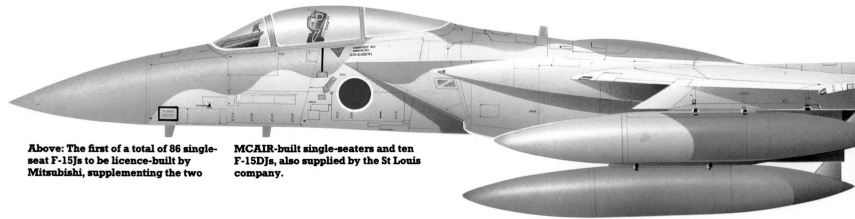

Above: The first of a total of 86 single-seat F-15Js to be licence-built by Mitsubishi, supplementing the two

MCAIR-built single-seaters and ten F-15DJs, also supplied by the St Louis company.

Above: One of the 15 F-15Ds which
Saudi Arabia has bought along with
47 single-seaters.

Above: Israeli deployment is
shrouded in secrecy, but it is known
that the country's initial order for 25
F-15A and B Eagles went to equip No
133 Squadron of the IDF-AF.

Below: One of the two MCAIR-built
F-15J Eagles arrives at Gifu air base
in Japan after a ferry flight from St
Louis.

'Foxbat' which had tried to intercept an Israeli RF-4E.

On Sunday June 7, 1981 the Israeli Air Force were busy again, this time some of their F-15s provided a CAP for F-16s that successfully bombed an Iraqi nuclear reactor nearing completion in Baghdad, thereby denying that nation the ability to manufacture nuclear weapons, at least for the time being. A month later on July 29, an F-15 bagged another 'Foxbat' which was endeavouring to intercept yet another IDFAF RF-4E.

During the Israeli invasion of southern Lebanon on June 6, 1982, their F-15s were in the thick of it, accounting for no fewer than 42 Syrian aircraft by the time the dust had settled. On August 31, 1982, an Israeli Eagle downed yet another 'Foxbat'.

On September 24, 1980 a simmering border war between Iraq and Iran flared into full-scale hostilities when Saddam Hussein's troops and tanks crashed across the Iranian border in a dawn thrust to encircle the world's largest oil refinery at Abadan. This was the beginning of a bitter, bloody war, which would cost hundreds of thousands of lives, and solve absolutely nothing when it was finally agreed to call a halt to the carnage eight years later.

During this conflict on June 5, 1984 two RSAF F-15s from 13 Squadron, based at Dharan, were vectored by a USAF E-3 Sentry on patrol over the Persian Gulf, to intercept two Iranian fighters violating Saudi airspace. Both F-15s launched AIM-7s and were able to claim a kill each – the intruders were discovered to be F-4E Phantoms.

The Iran-Iraq war left the economies of both countries in tatters. By July 17, 1990 the region was once again becoming highly volatile with Saddam Hussein this time accusing other Gulf states of conspiring with the US to cut oil prices and of "stabbing Iraq in the back with a poisoned dagger". On August 2 Iraq invaded its tiny neighbour, Kuwait; UN Security Council Resolution 660 condemned the invasion and demanded that Iraq withdraw its forces immediately and unconditionally. Five days later President George Bush sent 4,000 troops of the 82nd Airborne Division and F-15Cs of the 1st TFW, from Langley AFB, to Dharan, Saudi Arabia. In addition many other nations sent troops, ships and aircraft in support of Operation Desert Shield. The build up of allied forces continued relentlessly, more F-15 units were deployed to the region; air elements of the 36 and the 4 TFWs went initially to Tabuk, and formed the 4th TFW (Provisional-P). On August 27 and 28, 24 F-15Cs from the 58 TFS, 33 TFW based at Elgin AFB, were deployed to Tabuk Air Base, Saudi Arabia, where they formed the 33 TFW (P). Air activity and training intensified as units flew hundreds of sorties preparing for a war that all hoped Saddam Hussein would avoid. At 9.35am (local) on September 30, 1990, while participating in a low-level training sortie, an F-15E of the 4th TFW crashed in Saudi-Arabia, killing both the pilot Maj. Peter S. Hook and the WSO Capt. James B. Poulet.

Second Front

In December and January a second front was established at Incirlik AB, Turkey; assigned to the 7440 Combat Wing (Provisional) it consisted of more 36 TFW F-15s and the 32nd TFS from Soesterberg. In the Oval Office, President Bush signed a secret National Security Directive on January 15, 1991 which constituted a decision to enforce the 12 UN resolutions against Iraq, this cleared the way for a follow-on execution order to be passed through the National Command Authority – the Secretary of Defence, the Joint Chiefs of Staff and the military chain of command. This was signed by Secretary of Defence Richard Cheney and after the January 15 deadline had passed Cheney affixed a pre-agreed date to the document. The best efforts of the world's diplomatic corps had failed, Hussein refused to vacate Kuwait. During the evening of the 17th, President Bush broadcast to the American nation, "As I report to you, air attacks are under way against military targets in Iraq" . . . Operation Desert Storm had begun three hours earlier.

During the night of January 16/17, Capt. Steve Tate of the 71st TFS, 1st TFW was commanding a flight of four F-15Cs, providing a Combat Air Patrol (CAP) for a force of F-15Es, F-4Gs, EF and F-111s attacking an airfield SE of Baghdad. Two hours into the sortie, an E-3 Sentry reported a 'boggie' closing on number 3 in the formation. Tate flying 83-0017, call-sign Quaker 1, established that the aircraft was hostile via its IFF and at 12 miles obtained a lock. Closing in, he launched an AIM-7 from a range of 4 miles, which completely destroyed the Iraqi Mirage F-1; the time was 0001 Zulu (Z-G.M.T.), this was the first air kill of the war and the first of many for F-15s.

Above: Between sorties during Operation Desert Storm, ground crews work on F-15s on an airfield in the Saudi Arabian desert.

Elements of the 33rd TFS (P) were also airborne that night, hooked up to tankers, they were briefed to provide cover for returning bombers. However, thirty minutes before the time to leave, the tankers were alerted by an E-3 that the strike force was under attack from Iraqi fighters. They broke off and headed north, however as Capt. Larry

"Cherry" Pitts recalls, "our radar scopes were filled with friendlies – 60 to 80 of them! Night conditions combined with bad weather made it difficult to fire missiles even if the F-15s acquired targets. There were just too many friendlies out there". On entering Iraq, the formation broke into two four-ship flights. Capt. Pitts, "The whole ground was red with Triple-A fire as far as you could see in the area around Baghdad". Captains Tollini – flight lead and Pitts targeted several enemy aircraft but were unable to fire before the enemy fled.

Capt. Jon B. Kelk however, had better luck, flying 85-0125 he downed the 33rd's first enemy aircraft of the war, a MiG-29, at 0315 local, with an AIM-7. Just five minutes later Capt. Robert E. Grater flying 85-0105, in the other 33 TFW formation chalked up the first double kill of the war, when he destroyed two Mirage F-1s, again using AIM-7 missiles. The two

Below: Indicative of the harsh conditions of the region, two F-15s of 33rd TFW patrol with a RSAF F-5 through a mountain region.

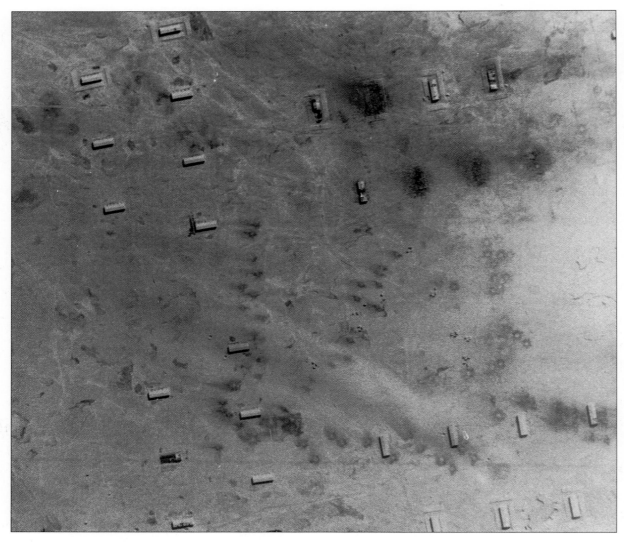

formations then rejoined over Iraq and returned to Tabuk.

Another eight-ship from the 33rd was airborne again that afternoon. Under the leadership of Capt. Charles "Sly" Magill – a U.S. Marine Corps (USMC) pilot, their mission was to sweep Iraqi skies clear of enemy aircraft ahead of a strike force of some 60 F-15s, F-16s and EF-111s, that were to bomb an airfield near Baghdad.

Whilst tanking, an E-3 reported two MiG-29s in the area, south of the strike forces target. Chuck Magill in 85-0107 split his flight into two four-ships, directing one to continue to sweep its pre-arranged airspace, he led the other aircraft which consisted of his wingman, 1st Lt. Mark J. Arriola, and Capts Rhory R. Draeger in 85-0108 and Anthony "Kims" Schiavi – after the MiGs. They came under heavy surface-to-air fire. Jettisoning their wing tanks they dived to lower altitude. Clearing the threat, it became apparent to the F-15 pilots that the Fulcrums were flying a low, quite slow north/south pattern to protect the airfield designated to shortly be attacked.

As the Eagles approached, the MiGs turned north away from them. The F-15s opened up the taps and began closing. The MiG-29s then turned south towards the Eagles, which brought them within the AIM-7 envelope. Capt. Draeger recalls, "I shoot first and Sly shoots right after I do. I could see the missiles from the time they left the aircraft going right to where they were supposed to. I could see mine, and could see Sly's trailing about two or three thousand feet back. They hit the two planes nose on". Both aircraft exploded into two fireballs. This was to be the only kill of the war credited to a USMC pilot, but the battle was not quite over. Now low on fuel and well to the north, the formation were again being targeted by Iraqi SAMs. They climbed back to altitude and jettisoned their last external fuel tank. Successfully performing a series of evasive

manoeuvres, the flight managed to elude the threat and RV with the tanker, after which they returned to Tabuk.

In addition to the blows suffered by the Iraqi Air Force, Command, Control, Communication and Intelligence (C³I) centres in Baghdad, Kirkuk, Rutbar, and Nasirif, as well as Nuclear, Chemical and Biological weapons producing facilities and Scud missile sites were all hit hard early in the campaign. The fixed Scud sites were easy meat for cruise missiles, the mobile Scud launchers were however, a different proposition.

Scud busting
The first of several Scud attacks occurred at about 0130Z on January 18, when approximately ten missiles were launched. At least one Scud was fired against Saudi Arabia and was intercepted and destroyed near Dharan by a US Army Patriot air defence missile. Seven Scuds were launched against Israel, two struck residential areas in Tel Aviv causing injuries to seven people, another struck in Haifa. The US moved swiftly to foil Hussein's crudely disguised attempt at provoking Israel into a military response that could split the allied alliance line-up against him, by supplying Patriot missiles and crews to help defend Israel in return for a non-intervention commitment. In addition an elaborate plan incorporating a Lacrosse radar imaging satellite, four KH-11s and two Air Force Space Command Defence Programme Missile Warning Satellites, made invaluable contributions to detect the mobile Scud launched menace. Other intelligence gathering assets included the E-3 Sentry and the only two E-8s in existence. Modified Boeing 707s, the Grumman E-8s were deployed in January and made-up the 441th Joint Surveillance Target Attack Radar Systems (Joint-STARS); flying 12 hours shifts to ensure 24 hour coverage, their high powered SAR could detect tracked

and wheeled vehicles at ranges in excess of 60 miles (100km). Detailed information concerning location etc. was then sent over its jam proof communications system to be correlated and tasked out to 'the real Scud busters'. Leading these night attacks were F-15Es of the 336 TFS which arrived in Saudi Arabia in August and the 335 TFS which arrived in the second half of December. "We listen for radio calls from controllers who will tell you quickly if any missile has been launched and who give you the co-ordinates very quickly. If they have launched a Scud, they get to pay the price if we find them – and we've had a fairly reasonable success rate", said Lt. Col. Bob Gruver, Ops Officer for the 336 TFS, during an interview with Jeffrey Lenorovita of *Aviation Week & Space Technology*. At the heart of this 'reasonable success rate', is the APG-70 radar and the LANTIRN system described earlier. The 335th were equipped with both the navigation and the targeting pod from the outset of hostilities and were therefore able to use highly accurate, precision-guided munitions; the 336th however had the targeting pod delivered and were checked out on its use as the war progressed – being forced to use MK. 82 'dumb bombs', CBU-52, 58s and 87s as well as MK. 20 Rockeyes, as an interim measure. The two units flew an average combined total of 40-60 sorties per night. Such a staggeringly high sortie rate, by an aircraft still undergoing operational test and evaluation was an outstanding achievement. To help solve problems encountered by 'E' models in the field, a programme called 'Desert Eagle' was implemented. A team from Wright-Patterson AFB, were sent to the 4th TFW (P) and relayed messages to the SPO, of squawks that came to light; technical experts Stateside were then able to rectify design deficiencies. Problems regarding sand in BRU-46 bomb rack release mechanisms, sand damage

to windscreens, weapon system certification and additional ground cooling for avionics, were all solved in record-breaking time by this system.

These activities were not however without losses, and resulted in the only F-15 casualties of the war. The first occurred on the night of January 17, when Major Thomas E. Koritz and his WSO Lt. Col. Donnia P. Holland, both from the 335 TFS "Chiefs", were shot down and killed while flying F-15E, 88-1689.

At 1600Z (20.00 Local) on January 19 twelve F-15Es from the 335 TFS got airborne for their base in Saudi Arabia. They were accompanied by another 12 E models from their sister squadron, the 336 TFS. Electronic warfare support was provided by two EF-111 Ravens and SAM suppression was supposed to have been supplied by F-4Gs. Flying 88-1692 was the 336 TFS Deputy Commander for Operations Col. David W. Eberly. His WSO was Major Thomas E. Griffith from the 335 TFS, this was Tom's third combat sortie and his first with Dave Eberly.

F-15 downed
The target was a SCUD missile storage area in western Iraq, not far from the Syrian border – it was from this region that SCUDs were being fired at Israel. Their aircraft, callsign CORVETTE 03, was armed with CBU-52-cluster bomb units and two AIM-9s for self defence.

To minimise the threat from 23mm and 37mm AAA (Anti Aircraft Artillery), the attack was to take place at 20,000ft and all ordnance was to be dropped on the first pass – known in the trade as "one pass haul ass".

About 2½ hours into the mission, with the SAR slewed to the side and a radar map of the target area showing clearly on the CRT, Corvette 03 turned into the target to make its run. Heavy AAA was by now snaking its poison trail up towards the attackers, but of more concern to Tom was the RHAW display, indicating a SAM launch. Now, with just 8 miles (12km) to run before release, Tom saw two fire balls coming up from the ground, he recalls "They were much larger and brighter than all the triple A and were clearly SAMs, coming up from the right. When it was clear that they had our names on them I put out some chaff and told Dave to come right. We were in a hard right turn when there was an explosion of bright light on our left". They were hit while travelling at 580kts indicated (420kts calibrated). Dave Eberly describes the impact "like the worst car wreck you could imagine in your life". Dave tugged the command eject handle and seconds later and quite miraculously, both crew members were floating under fully deployed parachute canopies. During the high-speed ejection Dave lost his helmet and suffered hypoxia on the descent.

Tom landed quite close to the wreck of '1692, but by establishing some common landmarks against the night sky, they were able to meet with one another. Day-break revealed a small hill about ¼ mile distant that would afford at least some cover in this dry barren region of scrub and rocks. They hid up during the day and on the night of the 20th, were able to use their survival radios to make contact with their squadron when they returned to give the target another pasting – up until that time, the fate of the two downed airmen wasn't known.

Deciding that they were too near a

defended area for any search and rescue (SAR) forces to extricate them and now almost out of water, they decided to walk west, towards the Syrian border. After covering about 4 miles, they discovered what appeared to be an abandoned building. They observed the place for sometime and then gingerly approached. An Iraqi soldier on the roof began shouting and loosed off a few rounds from his AK-47, this brought his colleagues tumbling out of the building and at 02.00hrs (local) on January 21, Dave and Tom were POWs.

They were taken at day-break to the regional commander's house, but that night, were moved to a prison in Baghdad. Tom was eventually released when hostilities had ended on March 3, Dave followed three days later.

At approximately 12.00 local, on January 19, Capts. Craig "Mole" Underhill – flight lead in 85-0122, Caesar Rodriguez in 85-0114, John Fischer and Pat Moylan, all from the 33rd TFW (P) were providing a CAP for an E-3, directing allied air assets over the war zone, when a formation of F-16s requested that they conduct a pre-strike sweep in the vicinity of an airfield that they were to bomb, in northwest Iraq. Happy to assist, they made an initial run at a group of MIG-29s

threatening the F-16s, but they turned and ran. As the Eagles flew further north they picked up a second set of contacts working to the west. Capt. Rodriguez recalls, "They started to run at us. AWACS called that they were thirteen miles out and we turned to engage them". Both Underhill and Rodriguez got a radar lock on the first aircraft, but weren't cleared to shoot. "By that time" recalls Rodriguez, "I was too close and reacted defensively to the south". However, shortly afterwards Underhill got confirmation from AWACS that they were hostile (MiG-29s) and destroyed one of the aircraft at a range of 5 miles with an AIM-7. Rodriguez then visually acquired the other MiG-29, which had been flying about four miles in trail of his former flight leader. Rodriguez, "We started the fight at a fairly high altitude. The bandit initially started a defensive reaction towards my wingman, but reversed his turn and allowed me to get inside his circle. At that point we were at medium altitude, and I began to manoeuvre in relation to the bandit to get into the weapons position. He then started a hard defensive turn down toward the ground and after about a 360 degree turn between the two of us, I was approaching the weapon engagement zone. He

decided to execute a last ditch manoeuvre which was essentially a split-s. As he reversed his turn and started to aim towards the ground, I realised that I couldn't make that turn. So I got away from him to get some turning room to catch him on the backside, when he ran out of room and impacted the dirt. I don't think he realised how low we had gotten".

That same day four F-15Cs were conducting a fighter sweep. Capt. Richard C. Tollini was lead, flying 85-0101, Capt. Larry "Cherry" Pitts was number two in 85-0099, followed by Capts. Jon Kelk and Mark Williams. At about 14.00 local an E-3 contacted Tollini with information concerning two MiG-25s flying in line abreast some 50 miles north and straight at the Eagles. As Tollini turned the formation north to engage the threat, two MiG-29s were detected by the E-3 also enroute for the F-15s from the northeast. The MiG-25s had slipped into lead-trail, about 10 miles apart. Capt. Tollini targeted the lead MiG, but his radar broke lock, and the formation temporarily lost contact with the boggies until they suddenly reappeared five miles (8km) in front, crossing the formation on the beam, and doing 700 knots at 500ft (150m). Both Williams and Pitts locked on

Above: The F-15E Strike Eagle was an early arrival to the Gulf theatre, and scored some notable successes particularly against Iraqi Scud sites.

at the same time, but with the targets now flying west to east, Pitts was in a better position to engage. He made a hard right turn, but because of its speed, the MiG disappeared off the outer limits of his radar. Capt. Pitts continues, "As I continued around, the MiG decided to make a 270 degree turn through south, west and back to north. As he turned the corner going north, he lost a couple of hundred knots of airspeed, and I was able to roll in behind him about 1½ miles away with a radar lock". While Tollini provided top cover against the other MiG, Pitts began firing missiles. He recalls, "He decoyed the AIM-9 with flares. I saw the pilot eject, but don't know if it was a successful one". Pitts then caught sight of the other MiG, and relayed its position to Tollini, who promptly acquired and despatched it with an AIM-7.

Below: Over flown by a C-5 transport, these F-15Es of the 4th TFW sit menacingly behind blast walls on an airfield in Saudi Arabia.

RSAF successes

The 36th TFW chalked up their first kills of the war at 1913Z that day when Capt. David S. Prather, flying 79-0069 and lst Lt. David G. Sveden in 79-0021 downed a Mirage F-1 each using AIM-7s. On January 24 a USAF E-3 detected two Iraqi Mirage F-1s flying at high speed and low level, south along the Persian Gulf coast. Capt. Ayhed Salah Al-Shamrani, was leading a flight of four RSAF F-15s on a CAP, when he was vectored to intercept the intruders that were 80 miles (128km) away and heading in the opposite direction. Ordering two of the four F-15s to remain on CAP, he and his wingman set off in pursuit. No attempt was made by the Iraqis to evade the F-15s, until the last minute, when one F-1 jettisoned its ordnance – believed to have been an Exocet anti-ship missile. They then commenced a left bank but

the AIM-9s fired by Shamrani slammed into their targets, destroying both aircraft.

Having a nose for trouble, the 58 TFS "Gorillas" were at it again on January 26. Four aircraft, Capt. Rhory "Hoser" Draeger, flight lead in 85-0119, together with Capt. Tony "Kimo" Schiavi flying number two, in 85-0104, Caeser Rodriguez, number three in 85-0114 and Bruce "Roto" Till, number four were providing high-value asset protection – to Tankers and an E-3. Shortly after tanking, an E-3 reported to Draeger that they had detected bandits some 120 miles away. As there was nothing else in the area, Draeger immediately committed his flight and they were off, in hot pursuit. After chasing them and making little progress, he was contemplating turning back, when the E-3 reported four more contacts coming from the same airfield

as those being chased. They too were flying north east. "Hoser" Draeger recalls, "The airfield was directly west of us, which put us in a perfect set up for an intercept. It was an overcast day, and we couldn't visually see the bandits, but our radar indicated all four of them. As we approached to within about 40 miles, one of the aircraft unexplainably turned back to its airfield". Draeger allocated targets and from 30 miles (48km) out they punched off their wing tanks. Draeger fired first from several miles out. He continues, "About that time a big hole appeared in the clouds and the four F-15s swooped down through it and spotted three MiG-23 Floggers following a road. My AIM-7 hits the back end of the first one, and I thought he was down right away because of the tremendous blast. All three jets were flying extremely low, and dust kicked up on him. I

Above: Flying in formation, these F-15s refuel from a tanker aircraft over the Gulf war zone during the Kuwait conflict.

thought he was in the dirt right away! But he keeps flying and the fire burns up towards the middle of the aircraft. Finally it blows up, the wings fly off and it does a quarter turn into the dirt". About two seconds later, the AIM-7 fired by "Kimo" Schiavi slammed into the second MiG. "Rico" Rodriguez's missile downed the third Flogger a fraction of a second before "Roto" Tills'. The four "Gorillas" then vacated the area, tanked and returned to Tabuk.

Below: Taken in February 1991, after an air strike, this reconnaissance photograph is of an Iraqi chemical factory in the Al Qaim region.

Right: Operating out of an air base in the featureless desert wastes, these F-15s prepare to take off on a sortie over Iraqi-held territory.

On the Iraq-Iran border

On January 29, four F-15s were sweeping the Iran-Iraq border to prevent Iraqi aircraft from escaping to Iran. Assigned during the war to the 58 TFS, back State-side the four pilots were all part of the 60 TFS "Crows". Capt. Mike Fischer was lead, Capts. Pat Moylan number two, Dave "Logger" Rose in 85-0102, three and Kev. Gallagher four. Arriving on station to relieve another 58 TFS four-ship that had covered the 6 AM 'til noon slot, the mission was quiet. The sky was overcast with cloud that extended from 16,000 to 32,000ft (4,800 to 9,600m). They chose to fly beneath it, in order to avoid any SAM sites in the area. Fischer and Moylan were tanking when at approximately 16.00 local Kev. Gallagher detected what he believed to be a hostile aircraft on his radar, about 60 miles (96km) away. Confirming via his IFF, that the contacts were indeed hostile "Logger" Rose and Gallagher set off to intercept them. At the time the airspace was quite crowded with F-15s from the 36 TFW to the north and a four-ship of Eagles from the 1st TFW to the south. Dave Rose therefore elected not to fire his missile BVR, but to obtain visual identification of the targets first. Manoeuvring into the enemy's 6 o'clock the Floggers were maintaining 1,000ft (300m) and more than 700kts. Rose remembers, "They were flying very low over the desert floor. I fired a missile at one. He blew up and hit the ground". Rose pulled off the target to allow his wingman a shot at the other aircraft. "As I turned east, I saw an F-15 from Bitburg also shooting at the MiG". On this occasion the MiG was extremely lucky as both aircraft missed their mark. Forced to leave the area through lack of fuel, they climbed to 42,000ft (126,000m) tanked and patrolled northwest of Baghdad until returning home.

Fourteen days after the most powerful conventional air campaign had been unleashed and just 100 hours after the land battle 'Desert Sabre' was launched; it was all over. A ceasefire agreement was signed on March 3, 1991, Hussein's forces were out of Kuwait (after having committed the most wanton acts of state-sponsored terrorism and vandalism that the world has witnessed for 50 years).

After the ceasefire

An uneasy peace followed, during which many Iraqis – including Sherias and Kurds, rose up and tried to rid the country of its loathsome dictator. At 1050Z on March 20, an E-3 detected two Su-22s above Takrit (Hussein's home town, about 100 miles (160km) north of Baghdad), flying in violation of the March 3rd ceasefire, prohibiting the flight of Iraqi fixed-wing aircraft, two F-15Cs of the 36th TFW engaged the aircraft and in the ensuing fight Capt. John T. Donski of the 22nd TFS, fired an AIM-9 from 84-0014 destroying one of the Iraqi jets – the other made a hurried landing. Virtually the same scene was re-enacted two days later with aircraft from the 53rd TFS. Capt. Thomas N. Dietz was flying 84-0010. At 1840Z Dietz despatched an Su-22 using an AIM-9. 1st Lt. Robert Hehemann was also able to claim a PC-9 Iraqi trainer which was flying in close vicinity of the Su-22, when its pilot decided he did not like the odds and ejected without Bob firing a shot.

In April 1991 the Department of the Air Force issued a White Paper, entitled *Air Force Performance in Desert Storm*. This is what it said about the F-15. "During Desert Shield, F-15s provided

the defensive umbrella that permitted the deployment of air, land and sea assets into AOR. After D-Day, they shifted to offensive counter air attacks against Iraqi Air Force and helped gain air supremacy within the first ten days of the war. Every Iraqi fixed wing aircraft destroyed in air-to-air combat by the Air Force was a "kill" for the Eagle. Their success permitted coalition air forces to exploit the versatility of airpower over the entire battlefield. The 120 F-15 C/Ds deployed to the Gulf flew over 5,900 sorties and maintained a 94% mission capable rate – 8% higher than in peacetime".

Of the F-15E the same document had this to say. "Forty eight of these multi-role fighters were deployed to the Gulf. The F-15E's flexibility was the key to its success. The F-15E proved its versatility by hunting SCUD missiles at night, employing laser systems to hit hard targets and attack armoured vehicles, tanks and

Above: The presence of the F-15 Eagle in the Gulf, such as this 1st TFW machine, was to prove a major asset to Coalition forces.

artillery. It proved unusually effective with the Joint Surveillance Target and Attack Radar System (JSTARS) for cueing on SCUD locations and using Low-Altitude Navigation and targeting Infra-red for Night System (LANTIRN) to locate and destroy the missiles and launchers. Its overall mission capable rate was 95.9% – 8% higher than in peacetime. These aircraft deployed with LANTIRN navigation pods (permits accurate navigation at night across featureless terrain to the target without the need of active navigation aids). Sub-

Below: The F-15s of the Royal Saudi Air Force played a major role in the great Coalition success over Iraqi and Kuwait air space.

sequently the targeting pods were deployed. During Desert Storm, the F-15E accomplished Operational Test and Evaluation of the LANTIRN system with spectacular results. Their primary targets were SCUDs, command and control links, armour, airfields and road interdiction. While over 2,200 sorties were flown, only two aircraft were lost in combat."

It is clear than in the F-15, MCAIR have designed an aircraft that will surely rank with other such classics as the P-51, F-86 and F-4. As mentioned earlier MISP will modify 300 F-15C/Ds and 150 A/Bs to include the AMRAAM system, Tracor ALE-45 chaff dispenser, Northrop ALQ-135 internal countermeasures system and the Loral ALR-560C RWR. These will ensure that the F-15 remains on the cutting edge of technology until the Advanced Tactical Fighter (ATF) enters service, around 2001, nearly 30 years after the first F-15 took to the air!

Performance and Handling

One of the fundamental aims of the FX programme was to produce a fighter capable of out-performing any actual or potential opponent, though there was considerable debate over the precise details of how this was to be defined. With its unprecedentedly high thrust-to-weight ratio and low wing loading for maximum manoeuvrability, the Eagle has demonstrated the required level of performance, and its handling qualities at speeds ranging from 100 knots to Mach 2.5 have won unqualified praise from pilots as well as providing a thrilling spectacle for air show audiences.

The performance requirements embodied in the FX Request for Proposals were intended to equip the resulting fighter to better all existing and projected Warsaw Pact opponents. As of the late 1960s these were represented by the small, agile MiG-21 and the rather bigger but highly manoeuvrable MiG-23, expected to be encountered at all altitudes; the fast, high-altitude MiG-25; and the Su-15 interceptor, for which the look-down radar capability was likely to be needed.

At the same time, some compromises had to be accepted, so that the USAF's original requirement for a maximum speed of Mach 2.7 at high altitudes, for example, was reduced to Mach 2.3, with a Mach 2.5 minimum burst capability. The higher speed would not only have prevented the use of a bubble canopy, added up to 3,000lb (1,360kg) to gross weight and reduced dash radius, it would also use up fuel at a rate of 65,000lb/hr (29,480kg/hr), equivalent to consuming the entire internal fuel load of an F-15A in about 11 minutes. The fuel flow at Mach 2.3 was estimated at a significantly lower 45,000lb/hr (20,400kg/hr). It was also argued that

higher speeds were in any case irrelevant to most missions.

Other performance figures specified included a top speed of Mach 1.2 at sea level, which it was felt would provide a useful margin of superiority over potential opponents. The wing was to be optimized for buffet-free performance at Mach 0.9 and 30,000ft (9,100m). More generally, performance was to be a consequence of high thrust-to-weight ratio and low wing loading, which had been recognized as the fundamental elements in providing the desired degree of superiority in performance and agility.

Wing loading

In its production form, at a takeoff weight of 41,500lb (18,820kg) with full internal fuel, the F-15A's 608sq ft (56.5sq m) of wing area gives a loading figure of just over 68lb/sq ft (333kg/sq m), and with half internal fuel this figure falls to 57lb/sq ft (279kg/sq m). Similarly, thrust-to-weight ratio at takeoff, with the F100s in full afterburner and full internal fuel, is 1.15:1, and by the time half the fuel has been used the ratio increases to nearly 1.4:1. By comparison, wing load-

ing of the F-4E at combat weight is 80lb/sq ft (390kg/sq m), with a thrust-to-weight ratio of approximately 0.85:1.

When translated into actual flying qualities these figures have important consequences. The four forces acting on an aircraft in flight are thrust, drag, weight and lift, and all are inter-related. Lift and drag increase as airspeed rises; both lift and drag are increased by increasing the angle of attack, and increasing the angle of bank in a turn also increases drag. Lift is considered as acting perpendicularly to the surface of a wing, and in a steep turn it has to counteract both the weight of the aircraft and the centrifugal force acting on it. Lift can be increased by increasing the angle of attack, but this also results in more drag, which in turn demands more power if the turn is to be sustained.

It is in this context that the theories of former fighter instructor Major John Boyd assume such importance. By the time Boyd was assigned to the FX programme in October 1966 he had already developed his theory of energy manoeuvrability in conjunction with Tom Christie, a mathematician working

at Eglin AFB. By subtracting drag from thrust and multiplying the residue by velocity, Boyd realized, it was possible to express the 'energy rate': when drag exceeds thrust the energy rate becomes negative, so that either more thrust must be made available or the aircraft will lose altitude, airspeed or both. This in turn gives rise to the concept of specific excess power (Ps), or the amount of 'spare' thrust available in a turn, and explains the importance of the F-15's unprecedentedly high thrust-to-weight ratio.

Wing loading also has an important effect on turning performance. As an aircraft banks, the amount of lift needed to counteract the combination of gravity and centrifugal force increases as the bank angle increases; and since the

radius of turn at a given airspeed depends on the bank angle, it follows that the turn radius of which an aircraft is capable is dependent on the extent to which it can continue to develop lift with increasing bank angles. The high available thrust enables the Eagle to maintain or increase speeds in high banked turns, thus enabling it to fly turns at high rates. Thus the F-15's low wing loading and high thrust-to-weight ratio combine to make it exceptionally manoeuvrable. Compared with the F-4, the F-15 can take off in a shorter distance, accelerate faster to a higher maximum speed, turn with a reduced radius and at a higher

rate and fly higher; alternatively, it can climb at a lower airspeed.

To allow pilots to make maximum use of the Eagle's power and agility a dual flight control system was developed. The conventional hydromechanical system operates through push rod linkages acting on the valves of hydraulic actuators which deflect the control surfaces. The pitch-roll control assembly is a mechanical system which modifies the response of the system and the aileron-rudder interconnect couples the rudders and stabilators so that the former are operated automatically in conjunction with the stabilators, allowing ma-

noeuvres to be carried out using the stick alone.

Meanwhile, the automatic control augmentation system (CAS) forms a separate fly-by-wire system using electrical signal signals and servo motors to operate the hydraulic actuators. The CAS system includes pitch and yaw rate, angle of attack and dynamic pressure sensors, as well as accelerometers to monitor vertical and lateral acceleration. It is thus able to compute the correct settings for the control surfaces at any combination of speed and g forces.

The CAS also senses the stick forces

Above: A 36th TFW F-15A in formation with an F-5E in the blue-grey camouflage of an Aggressor squadron. The Aggressors simulate MiG-21s for combat training.

applied by the pilot, translating them into electrical signals to the control surface actuators; should the mechanical system fail the CAS would continue to operate the control surfaces. For

Below: Even with a full load of three external tanks, the F-15 is capable of impressive takeoff performance and rapid climb.

Above: Three 405th TTW F-15As in a steep climb over the Arizona desert with engines in full afterburner delivering 11 tons of thrust each.

Right: Performance envelope of the Strike Eagle ground-attack version of the F-15 remains impressive even with external stores.

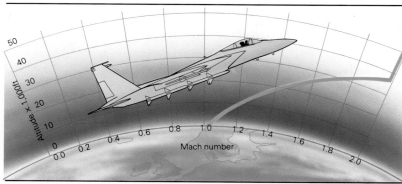

safety, the CAS is a dual-channel system in which the signals carried by each channel are continuously compared with each other, and if an error greater than a predetermined maximum is detected both are automatically disengaged. The Eagle can be manoeuvred with the CAS off using the mechanical system alone: in this mode, control is reportedly equal to that of earlier fighters, with the CAS providing a marked improvement in handling and effectively intervening to point the aircraft in whatever direction the stick is moved.

The control surfaces themselves are straightforward. The all-moving stabilators act in unison for pitch control and differentially for roll control in conjunction with the ailerons. Being mounted outside the engines, the stabilators could be positioned out of the wing wake without interference from the jet efflux, and at high angles of attack, when the ailerons become progressively less effective, the tail surfaces generate all required roll moments. The twin vertical stabilizers are tall enough to maintain stability at high angles of attack, and the twin booms on which the tail surfaces are mounted are braced to each other via the titanium engine bay assembly to provide a rigid structure for transferring the torsional loads from the tail to the aircraft.

Modifications to the aerodynamic configuration and control system as a result of flight testing are described in detail in the first chapter. Structural alterations included the clipping of the wingtips to reduce buffeting at high subsonic speeds and high g at around 30,000ft (9,100m); the increased chord on the outer sections of the stabilators to eliminate flutter, producing the notched leading edges of these surfaces; and the doubling in size of the speedbrake.

At the same time, several changes were made to the flight control system, such as the reduction in stick control forces during high-g manoeuvring to make CAS-off control easier. The CAS itself was modified to be less responsive to small, sharp stick deflections, since its original bias towards rapid rolling made the aircraft's response alarmingly jerky during more precise manoeuvres such as formation flying, air-to-air refuelling and target tracking with the gun. Similarly, the aileron-rudder interconnect system was made to disconnect on touch-down to eliminate the accentu-

Right: The F-15 has demonstrated an ability to turn well inside an F-4E, or climb 7,100ft (2,164m) while matching the Phantom's tightest level turn.

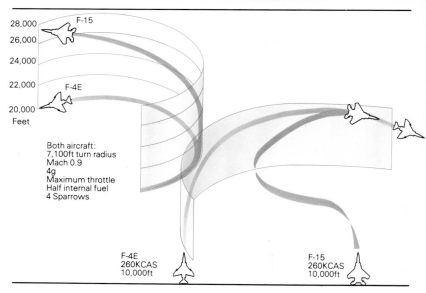

Both aircraft:
7,100ft turn radius
Mach 0.9
4g
Maximum throttle
Half internal fuel
4 Sparrows

Above: Superb performance is matched by a magnificent view from the cockpit: a 49th TFW F-15A is in the background.

Left: A pair of 57th FWW Eagles demonstrate the type's excellent low-level handling qualities over Nellis AFB, Arizona.

ation of weathervaning during cross-wind landings.

While all these problems were being identified and corrected during the flight test programme, the new fighter was demonstrated to a number of journalists during 1974. Clark Martin, of the US journal *Aviation Week and Space Technology*, flew in an F-15B when the aircraft was still subject to a limit of +5.9g: the left afterburner failed to ignite until after liftoff, but the takeoff was still impressive, a 3,300ft (1,000m) ground roll taking 18 seconds to a liftoff speed of 175kt. This compares with a ground roll of only 900ft (274m) for a production single-seater in standard interceptor configuration, when the nose-wheel is rotated at around 100kt and liftoff takes place at 135kt.

Following liftoff the F-15B went into a 50deg climb at 300kt, reaching a height of 14,000ft (4,267m) by the time it was over the end of the 15,000ft (4,572m) runway. During subsequent manoeuvres Clark reported that CAS-off flight was equal to that of earlier fighters, enabling tracking manoeuvres at up to 4g to be carried out at 10,000ft (3,000m) and 350kt. Climb and acceleration were also demonstrated, with a climb at better than 6,000ft/min (1,800m/min) in Military (non-afterburning) power to 32,000ft (9,750m) followed by

acceleration in full afterburner from Mach 0.9 to Mach 1.1 in 10 seconds and Mach 1.2 in 20 seconds.

Another close observer of an early demonstration flight was Captain Robert J. Hoag, Editor of the USAF *Fighter Weapons Review*. Following in an F-4 chase plane, Captain Hoag watched the Eagle lift off after a ground roll of 1,200ft (366m) and reach 10,000ft (3,000m) over a ground track of less than 5,000ft (1,500m). To compare the turning performance of the two aircraft, the F-4 initiated a 5½g turn in full afterburner, starting from a speed of 350-400kt and at an altitude of 12,000ft (3,650m). The Eagle, starting at the same speed and altitude and at a slant range of 6,000ft (1,830m) was able to close to a minimum range on the Phantom's tail within 540deg of turn, using only military power and with no trace of the severe buffeting experienced by the F-4.

Another manoeuvre which particularly impressed Captain Hoag, who was moved to describe the F-15 as a "Superfighter", was a transition from 110kt flight, with landing gear and flaps down, to an Immelmann in full afterburner with gear and flaps up. The transition was immediate, and the Eagle was able to accelerate throughout the manoeuvre.

The implications for combat of the

55

Above: The low-level flying capability of the F-15 is caught perfectly here from the aft cockpit of the trailing aircraft.

power implicit in such demonstrations are obvious. While the Eagle should be able to accelerate out of trouble fairly easily, an opponent would find it extremely hard to get away. The F-15's ability to maintain controllable flight over a wide range of speeds and at angles of attack up to 26deg can enhance the effectiveness of both missiles and guns. And the ability to maintain acceleration at steep angles is a significant advantage against the high-speed, high-altitude MiG-25, allowing the target to be tracked by the radar against a clutter-free background in a snap-up intercept.

Overall, the Eagle's air combat superiority is directly attributable to the high lift and excess power specified in the original requirement. Whereas the pilot of an F-4 must be aware of the need to conserve energy, and will often have to build up energy in combat by diving away into a zoom climb, the F-15 has so much power available that it can accelerate straight into high-g manoeuvres, and sustain them for long periods without losing altitude or airspeed.

Of course, there are other requirements in combat than pure power and manoeuvrability: above all, the pilot must be able to use them effectively. In this respect the bubble canopy and the exceptional visibility it provides is of fundamental importance – according to Captain Hoag, "The pilot feels as if he is riding astride the bird, rather than in it". There is also room for him to twist around in his seat without banging his head against the canopy.

Cockpit layout

The cockpit layout itself was given an enthusiastic welcome. The most frequently used controls are positioned at the top of the centre instrument panel (HUD, UHF radio, Mode 3 IFF and position identification) or on the left console near the throttle (radar controls, fuel jettison, ILS TACAN, landing lights, BIT control panel auxiliary radio and master IFF). Other instruments on the front panel include the air-to-air and air-to-ground weapons monitoring systems, the radar display, standard flight instruments and engine instruments. The right console contains the engine controls, INS control panel and environmental controls. Altogether there are 30 per cent fewer gauges than in an F-4 cockpit.

The pilot is further helped to keep his head up by the location of the main weapon and radar controls on the stick and throttle. With the CAS translating the pilot's inputs into electrical signals, finger and thumb pressure on the stick is enough to maintain control, while the controls for a number of other functions are also carried on the stick. A trim button allows the aircraft's attitude to be adjusted in pitch or yaw axes; the trigger engages the HUD camera and fires the gun; the weapon release button, depending on the mode selected, will engage the HUD camera, launch a selected missile, release bombs, designate a target or illuminate a target for a Sparrow; another button uncages or cages the seeker heads of heat-seeking Sidewinders or laser-guided bombs; and the air refuelling receptacle re-

Right: Touchdown by a 555th TFS Eagle. Even without a brake parachute the F-15 needs a shorter landing run than the F-4.

lease switch, when the radar is engaged, enables the pilot to engage the radar's boresight or super-search modes. The former commands the radar to scan directly ahead and lock on the first target it detects within ten miles (16km), while in super-search the radar scans the 20deg HUD field of view up to ten miles ahead and locks on any target detected, displaying a box in the HUD to indicate its position to the pilot.

The throttles also carry a variety of controls. Apart from the microphone and IFF transmitter, there is an ECM dispenser switch for releasing chaff or flares, a weapon mode selector for the air-to-air radar mode appropriate to gun, Sparrow or Sidewinder, and the speedbrake control, plus more radar controls. The antenna elevation control allows the radar scan pattern to be adjusted up or down, while the target designator control adjusts the position of the radar scope target acquisition gate, used to designate a target detected in normal search modes.

HOTAS philosophy

The thinking behind the head-up HOTAS (hands on throttle and stick) system of combat is summed up succinctly by Colonel Wendell H. Shawler, Director of the F-15 Joint Test Force, 1973–76: "Everything you need is on the throttle and stick". However, he adds that these functions need to be simplified to avoid confusion during the heat of battle. According to Jack Cranes of MCAIR, addressing the Society of Experimental Test Pilots on the F/A-18 HOTAS system, the average pilot at any given time can only use three functions intelligently, wisely and properly. Air-to-air weapon seleciton on the F-15 is split into three functions: gun, Sidewinder or Sparrow.

Again according to Colonal Shawler, on selecting guns, "you automatically

Right: The pilot's controls of the F-15. Note the range of controls on the stick, and on the top of the picture the controls of the HUD.

got everything up there (in the HUD) you needed for guns, including short-range radar on your radar selection". The same applies to the two types of AAM. "The fact that you were selecting one of three functions – you got everything you needed. You couldn't do much else – you don't want to. When you start getting all these complexities where you put 50 functions on the stick and throttle, if you select one thing over here, you get three choices there, it gets very difficult." During Colonel Shawler's time HOTAS did not strain the pilot's workload; but later, when all the ECM and EW aspects were added to the one-man cockpit, it did, in his opinion, "put the workload over a lot of people's capabilities".

This view has been echoed by at least one experienced F-15 pilot, Major Dito Ladd, who until transferring to the 527th Aggressors was Chief of Weapons and Tactics with the 525th TFS, 36th TFW. In May 1983 Major Ladd recalled: "There are times when you feel like 'Yeah, I've got this thing whipped, I can take advantage of every opportunity the weapon provides.' That was generally after a period of intense flying. Flying once a day or twice a day over a period of two to three weeks there are times when the airplane can be fun, with one man it can be extremely effective, but it take the time and practice to do it. It's just like anything, you've got to practice."

Right: The pilot's work in the F-15 is eased a great deal by the layout which concentrates most of his energies into avionics and ordnance control.

Improving the Breed

Although intended purely as an air-superiority fighter, the Eagle has also proved a highly successful convert to the ground attack role, with the F-15E Strike Eagle more than proving its combat effectiveness during Operation Desert Storm. Now this versatile and powerful aircraft is being given further capabilities in the STOL/MTD programme.

To ensure that a fighter remains at the cutting edge of combat capability, continual upgrading of materials, powerplant and avionics is essential. Nowhere has this policy been more stringently applied than with the Eagle. Additional fuel tanks located in the fuselage and wing leading and trailing edges increased internal tankage by an extra 1,855lbs (841kg). The undercarriage was strengthened and a Fuel And Sensor Tactical (FAST) pack conformal fuel pallet, designed and built by MCAIR in just 139 days were fitted, one below each wing and adjacent to the side of the fuselage. These increased the Eagle's fuel capacity by a further 9,750lbs (4,422kg). They were first test flown on F-15B 71-291 on July 27, 1974 and despite being flight tested throughout the entire envelope, it was found that they didn't have a detrimental effect upon the aircraft's overall handling qualities – except in one out of control mode. Demonstrating the aircraft's strategic reach in this configuration, Irv Burrows (MCAIR's F-15 Chief Test Pilot) and Lt. Col. Wendall 'Wendy' Shawler (Chief USAF F-15 Test Pilot), flew 71-291 un-refuelled, across the Atlantic, from Loring AFB, Maine to RAF Bentwaters, Suffolk. It later flew from there to RAE (Royal Aircraft Establish-

ment) Farnborough and participated in the bi-annual trade show, before touring and demonstrating at several USAFE bases in the UK and Germany.

Further improvement came with Hughes pioneering development of a Programmable Signal Processor (PSP). Appearing first in 1979, it was the first PSP ever deployed, expanding the F-15's tactical air interdiction role, while enhancing air superiority capability by replacing the earlier 'hardwired' system with the new 96k unit. This high speed special purpose computer, which controls the radar modes through its software, allows rapid switching of modes for maximum operational flexibility. In addition new weapons systems can quickly be integrated into the aircraft's fire control system, simply by re-programming the PSP's software.

The first F-15C (aircraft number 446, serial 78-0468) which incorporated the above improvements, flew for the first time on February 27, 1979. The new two-seat F-15D (aircraft number 450, serial 78-0561) first flew on June 19, 1979.

Eagles employ the Lead Computing Optical Sight System (LCOSS), which requires the pilot to track the target long enough for the gunsight to use the turn rate of the F-15 to compute the lead

needed on the target. For the cost of an AIM-7 Sparrow a-piece, every F-15 in the fleet could have been equipped with an Integrated Fire/Flight Control (IFFC) System. Successfully evaluated and demonstrated on the F-15, IFFC allowed the aircraft's flight control system to complete the fine gun tracking solution and with the trigger already depressed, the weapons system automatically actuated itself when all engagement parameters were met. Alas, the system was never deployed operationally.

The MSIP

A two phase Multi-Stage Improvement Plan (MSIP) was implemented tasked with monitoring enhanced radar improved software and the introduction of JTIDS – the Joint Tactical Information Distribution System (JTIDS). In phase 1 of the plan, a 5 × 5in (12 × 12cm) Shadowmask, colour cathode ray tube (CRT), supplied by Sperry Flight Systems was installed. The unit fulfilled several functions including Built-in test (BIT) displays, graphic display of armament stores, video display of electro-optical sensor systems and weaponry display/interface, in readiness for the possible operational deployment of JTIDS.

Development of the PSP also paved

the way for the modification of an AN/APG-63 to provide a synthetic aperture radar (SAR) function. This followed on from earlier work in the USAF Forward-Looking Advanced Multi-mode Radar (FLAMR) Programme. The new radar was first flown on an F-15 in November 1980 and by the 40th test flight, target resolution was down to the stipulated level of 10ft (3m). This translates in practical terms to an aircraft the size of a Pitts Special being recognisable from two thirds of a mile (1km) square radar map of an airfield 10 miles (6km) from the F-15.

The real break-through however, came with the implementation of MSIP II and the deployment of a new radar, The Hughes AN/APG-70. In this powerful unit, the radar data processor memory was increased from 16K to 24K and its processing speed tripled to 1.4 million operations per second. Gate array tech-

Above: Dive-bombing demonstration by the second development F-15B, 71-0291, in its Strike Eagle configuration in 1980.

Below: The first of a new breed, the F-15B aircraft 71-291 in its demonstration livery banks over the city of St Louis.

nology enables the PSP to now operate 34.5 million operations per second – that's five times faster than earlier PSPs – and provides 10 times as much memory. The APG-70 BIT capability has 10 times the software and six times as many test targets as the APG-63, which has led to a 33% increase in Mean Time Between Failures (MTBF). Additionally the new unit boasts a multiple bandwidth for high resolution ground mapping using the SAR technology mentioned earlier, and an analog signal converter. Flexibility made possible by programmable software, means that all weapons systems carried by the F-15 are compatible with the APG-70.

In addition phase II of MSIP provided provision for a JTIDS terminal. This system designed and manufactured by Collins Government Avionics and Singer's Kearfott Division will enable Air Force and Army elements to co-ordinate missions with reliable, real-time information, by enabling data to be transmitted securely and automatically, using the existing Tacan antennas for both Tacan and JTIDS signal reception. (At the time of writing however, this system had not been deployed on the F-15 operational fleet). The first MSIP II aircraft to be rolled out and flown on June 20, 1985, was F-15C, aircraft number 908, serial 84-001.

Any modification that was likely to change the performance or handling characteristics of the aircraft in any way, had to be thoroughly checked-out in flight test. As the 'turkey-feathers' – the flaps that covered the engines exhaust nozzles, were lost at regular intervals, a Component Improvement Programme (CIP) test flight was undertaken by Major Tom Tilden – and AFFTC pilot at Edwards AFB, to determine if the F-15's performance was in anyway impeded with all the 'feathers' removed. Tom recalls, "One task was to start at Mach 1.2 and 10,000ft [3,000m], do a level 5g turn and decelerate to 0.8 Mach. I went to Mach 1.2 plus a little, lit the afterburner again and rolled into a level 5g turn.

Above: Command and status, tactical situation, radar, and duplicate of the pilot's HUD displays dominate the aft cockpit of the F-15 AFCD.

Surprise! The F-15 accelerated. The only way to reach the test point was to go below 0.8 Mach, light the burner, go into a 5g turn and accelerate through the speed of sound while in the 5g turn. For an ex F-4 pilot like me, this was awesome." Conclusion: removal of the

Below: The F-15 was the first aircraft to fly with the Jont Tactical Information Distribution System (JTIDS).

Above: The 'Strike Eagle', (F-15E) prototype, an F-15B serial 71-291, which first flew in its new configuration July 8, 1980.

Below: The F-100-PW-220 Turbofan engine undergoing a sea level test firing within a stand. The engine was introduced in 1985.

'turkey feathers' had little or no effect on the Eagle's performance!

Engine development continued with an F-15, aircraft number 9, serial 71-287 operated by NASA during an Air Force sponsored programme aimed at expanding the capabilities of the F-100 engine through use of single-crystal turbine airfoils, an advanced multi-zone augmentor, an increased airflow fan and a digital electronic engine control system. According to Pat Henry (by this time MCAIR, Director of Flight Test Ops) . . . "improvements achieved in engine acceleration and afterburner operation were dramatic. Tests included a side-by-side comparison with another F-15 powered by the standard P&W F100-PW-100. The two aircraft started simultaneously, and the F-15 with the digital control rapidly outdistanced the other aircraft". Another significant feature of the engine was its stall-free operation throughout the flight envelope, even during aggressive throttle transients.

From this programme was born the Pratt & Whitney F100-PW-220 engine, qualification of which was completed in March 1985. Production introduction occurred in November 1985 and Operational introduction seven months later. The emphasis is on reliability and durability. For the first time F-15 pilots could confidently slam both throttles from Mil to Max AB and 4 seconds later enjoy 24,000lbs (10,886kg) of thrust from each engine, without having to worry about possible engine stagnations. Such ease of operation is due primarily to the digital electronic engine control system. Other features however include a 4,000 cycle first inspection core, 1,200 hour augmentor and the 2,000 hour gear fuel pump – the latter being constructed from less than 60% fewer parts than the earlier model and has a predicted life three times longer. In addition Pratt & Whitney offer an F100-PW-220E retro fit kit that incorporates the digital electronic control, core components and gear-type fuel pump which effectively upgrades the earlier F100-PW-100 engine to −220 status at about one third the cost of a new −220 engine.

Strike Eagle

In a company sponsored programme called 'Strike Eagle', MCAIR took aircraft number 12, F-15B serial 71-291, enhanced its avionics suite, missionised the cockpits, and put a Weapons Systems Officer (WSO) in the back seat. The newly configured aircraft first flew on July 8, 1980. Rows of stub pylons on the lower corner and bottom of the FAST packs – now referred to as Conformal Fuel Tanks (CFTs) were flight tested, during which it was demonstrated that in this tangential carriage configuration, an F-15 could haul three external fuel tanks, 12 × 500lb (227kg) bombs, four air-to-air missiles and two infra red sensor pods, 40% further than with conventional weapon configurations. Indeed, earlier

war cries of 'not a pound for air-to-ground' were becoming fainter as the diverse abilities of the F-15 were being realised.

The aircraft participated in the September 1980 Farnborough Trade Show and was soon undertaking numerous and various flight trials.

Within TAC circles the need to replace the ageing F-4 and augment the heavily committed F-111 force in the interdiction role and for that matter, the equally heavily committed F-15 in the theatre air defence mission, was all too apparent. To meet this challenge, the Air Force undertook to analyse a dual role (air-to-air and air-to-surface) variant of the F-15 and F-16. In all some 400 dual role fighter would be needed – these would not be additional aircraft, but rather a realignment of requirements placed upon the planned fighter force as a whole.

The F-16E (E being the planned designation of the dual role version of the F-16XL) flight test programme was conducted at Edwards AFB, using two versions of the aircraft, each incorporating a characteristic cranked arrow wing which extended to replace the horizontal tail unit. Extensions both forward and aft were inserted in the fuselage for additional fuel. The undercarriage was beefed up to allow for the carriage of heavier stores, some of which were

carried on semi-conformed pylons. Aircraft number one, a single seat F-16 powered by a Pratt & Whitney F-100 engine flew a total of 205 flights, logging 246.6 flight hours. The second F-16, a two-seater powered by a General Electric F110 engine flew 182 flights in 195.5 flight hours. MCAIR fielded four F-15s. Aircraft number 641, F-15D serial 80-0055, a two-seater flew weapons separation tests over 22 sorties at Eglin AFB. F-15B – the flight demonstrator mentioned earlier, flew 67 sorties over 122.2 hours from Edwards during evaluation of the Synthetic Aperture Radar. Aircraft number 446, F-15C serial 78-0468 flew 91 sorties from Edwards AFB conducting fully instrumented performance and flying qualities evaluations and aircraft

number 763, F-15D serial 81-0063 completed 36 operational evaluation flights, from Edwards AFB by May 5, 1983.

Brig. Gen. Ronld W. Yates, Deputy for Tactical Systems, Aeronautical Systems Division, headed the Duel Role Fighter steering group which was responsible for managing the evaluation. After closely analysing the results, it fell to General Charles A. Gabriel, Air Force Chief of Staff to announce on Friday, February 24, 1984, that MCAIR had won the multimillion dollar contract.

Below: The MCAIR Strike Eagle in its original form, before conversion to AFCD and dual-role demonstrator for USAF trials.

Below: The F-15E on take-off. Major elements of the existing undercarriage were strengthened to allow the aircrft to carry heavier stores.

The F-15E

Aircraft number 986, F-15E serial 86-0183 undertook its maiden flight from MCAIR's St Louis plant, on December 11, 1986 with Gary Jennings at the controls. Today, five years later the aircraft has proved itself in combat and will rank amongst other such classic aircraft as the Mustang, F-86 and F-4. Re-stressed and strengthened to clear it for 16,000 hours – double the life of earlier F-15s – this 9g airframe is well capable of hauling a gross weight of 81,000lb (36,741kg) heavenward. A composite-structure wing statically tested back in 1974 was not pursued in keeping with the earlier low risk/cost development policy. However, the entire rear fuselage pod of the new fighter is manufactured from SPF/DB (superplastic formed and diffusion bonded) titanium. This has led to a neater, lighter more capacious engine bay, brought about by a saving in the number of separate components previously used. Reliable performance was achieved from day one with the two Pratt & Whitney F-100-PW-220 engines; the CFTs provided great range at a low drag coefficient which was further enhanced by tangental carriage of stores. Flight control actuators by National Water Lift, are driven by a triplex digital (fly-by-wire) system developed by Lear Siegler Astronics. This allows hands-off terrain – following at 200ft (60m), even after a single channel failure.

Co-ordinating radar data from its powerful system, an E-3 Sentry crew can direct F-15Cs maintaining a CAP (Combat Air Patrol) to strike at targets as they appear. Utilising their highly accurate Honeywell RLF (ring-laser gyro) inertial navigation system to feed position details to the moving-map display, the Eagle crew can set-off even during the darkest night.

Through the APG-70 the WSO has the ability to acquire from a single radar sweep an SAR picture even from low level, of airfields, in excess of 80 miles (130km) range. This picture can then be frozen, thereby reducing to a minimum 'give away' radar emissions. As the Eagle nears its target, a further radar sweep updates the picture in order to identify individual targets, such as tanks, missile sites or aircraft on an airfield, enabling the crew to fine tune their First Run Attack (FRA) plans. The WSO using his hand controls can then switch to his Low-Altitude Navigation and Targeting Infra-Red for Night (LANTIRN) system to press home the attack. This system comprises of two external pods, the navigation pod located under the right engine intake contains a day/night automatic terrain – following radar and a night FLIR (forward looking infra-red), presenting the pilot on his wide – FOV (field of view) HUD or the WSO on his CRTs, a forward picture of the terrain ahead. The targeting pod located under the left engine intake consists of a high resolution FLIR to provide day/night and bad weather stand-off targeting capability from about 10 miles (6km) range, a missile boresight correlator and a laser designator. The last two are for use in conjunction with Maverick (TV, IR or laser) missiles or GBUs (Guided Bomb Units, mainly using laser homing). To contend with enemy radar guidance systems the WSO has at his disposal an updated ALQ-135 system which now features a broadband jammer called Band 1.5. The

Above: The first F-15E under test by MCAIR. The aircraft is carrying prototype LANTIRN pods and conformal fuel tanks, which allow a marked extension in range.

Below: With a sophisticated system of avionics and a 24,500lb (11,113kg) ordnance load, the F-15E carries a hefty punch.

antenna for which is located at the trailing-edge root of the left horizontal stabiliser and leading edge roots of both wings. Band 3 transmitters are to be found in the inboard leading edges and the trailing-edge root of the right horizontal stabiliser – thereby dispensing with the blade antenna located below the nose of the A, B, C and D models. Having delivered all its dumb or smart bombs and pulled off the target, the F-15E still retains all the sub-sonic air-air fighting agility of its strictly air superiority forerunners.

MSIP III is a massive undertaking which calls for the updating of all earlier USAF Eagles, to F-15C standard. Over 160 E models will roll off the St. Louis production line. Conversion of earlier models based in the USA is taking place at Warner-Robins Air Logistics Centre, Georgia, USAFE aircraft are being converted by CASA at Getafe, Spain and PACAF F-15s are being sent to Kim Hue AB, South Korea.

The STOL/MTD Programme
On October 3, 1984, Aeronautical Systems Divisions Flight Dynamics Laboratory awarded a £117.8 million contract to MCAIR for an advanced development Short Take-Off and Landing (STOL), Manoeuvre Technology Demonstrator (MTD). Utilising F-15B, 71-290, the objective, according to Air Force programme manager Lt. Col. Richard A. Borowski was to demonstrate technology required to operate high performance fighter aircraft from bomb damaged airfields.

Test objectives required the aircraft to take-off carrying full internal fuel and 6,000lbs (2,700kg) of payload from a runway studded with bumps up to 4½in (11cm) high, then land at night and/or in bad weather without recourse to any external landing aids in 30kt cross-winds, at descent rates up to 12.5ft (3.75m) per second, in order to land on an area of runway just 1,500ft × 50ft (450 × 15m) – normal landing distance for an F-15 varies from 4,800ft (1,440m) to 6,000ft (1,800m).

Highly modified, the demonstrator features large movable canards on the forward fuselage to improve overall stability. To absorb the considerable extra punishment received during the heavy lands, Cleveland Pneumatic Corporation modified the main landing gear. The upper wing skins were removed and replaced by aluminium-lithium, which is just as strong, but 9% lighter than conventional aluminium. In addition an Integrated Flight/Propulsion Control

(IFPC) system was developed by MCAIR and produced by General Electrics Flight Control Division. Using a new computer chip, Ada and other higher order computer software languages, the system manages all control parameters. To minimise cockpit workload, the IFPC has five models of operation, conventional, short take-off/approach, short landing, cruise and combat. Position sensors linked to the throttle stick and rudder pedals feed electrical signals to the IFPC system. Control laws programmed into the flight controllers then analyse all inputs to determine what combination of the quadruplex fly-by-wire actuators need to be activated to execute precisely the manoeuvre required.

Phase one of the programme began at MCAIR's St. Louis facilities when at 11.54 on September 7, 1988, company test pilot Larry Walker flew the modified aircraft for the first time on a sortie lasting 1.2 hours. In all 43 test flights were conducted with the aircraft equipped with standard circular jet nozzles, after which Air Force and contractor personnel installed and ground-tested a pair of two-dimensional thrust-vectoring, thrust reversing rectangular nozzles. Built by Pratt & Whitney and incorporating the latest advances in both design and fabricating techniques, they are manufac-

tured from chemically-milled, welded titanium honeycomb, enabling the nozzles to operate at higher temperatures. These lighter units incorporate flat upper and lower flaps, driven independently to adjust the nozzle profile and/or vector thrust up or down, resulting in a full ± 20 degrees of movement.

First flight of the F-15 STOL/MTD in this configuration was on May 10, 1989, from Lambert St. Louis. After a further four flights the aircraft was ferried on June 15 to the Air Force Flight Test Centre (AFFTC), Edwards AFB. Seventeen flights conducted over June and July successfully demonstrated the thrust vectoring feature of the new nozzles and validated changes made to the IFPC systems software.

In December 1989, the F-15 STOL/MTD demonstrated its Autonomous Landing Guidance capability, by landing at Edwards AFB at night with no runway lights or ground-based navigation aids. Three months later flight testing resumed at Edwards and on March 23, 1990, the first in-flight operation of the thrust-reversing nozzles was conducted. Thrust-vectoring assisted take-off tests in April demonstrated a 25% reduction in take-off roll. Using thrust-reversing and anti-skid autobraking, another milestone in the programme was achieved when

Above: As was proved during the Gulf War of 1991, the faith placed in a ground-attack version of the F-15 was fully justified.

on May 22, the aircraft landed on just 1,650ft (495m) of runway.

At 09.42 hours on June 21, 1990, 71-0290, callsign STOL 12 got airborne from Edwards AFB; at the controls was Lt. Col. Greg Lewis, an AFFTC pilot, in the backseat, Larry Walker. Major Erwin 'Bud' Jenschke was flying chase in F-15A, '086, callsign Eagle 13. They climbed to 40,000ft (12,000m) and accelerated out to Mach 1.4. Flying loose line-abreast, both aircraft simultaneously undertook a rapid deceleration. The F-15 STOL/MTD used thrust-reversing for in-flight deceleration; Greg Lewis was at Mach 0.8 in just 30 seconds, it took Bud Jenschke a further 15 seconds to stabilise to the same speed, by which time he was over one mile ahead of STOL 12 – definitely not the place to be in combat!

The following fourteen months were taken up expanding the aircraft's thrust-vectoring and thrust-reversing envelopes and demonstrated how these technologies could be applied in combat. The programme ended on August 15, 1991, having comfortably accomplished all test objectives.

Right: The F-15 STOL/MTD prototype in flight. The thrust vectoring nozzles on the Pratt & Whitney engines can be clearly seen.

Glossary and abbreviations

AAA	Anti-aircraft artillery
AAM	Air-to-air missile
AAR	Air-to-air refuelling
AB	US air base (on allied territory)
ADF	Automatic direction finding (system)
ADTAC	Air Defense Tactical Air Command
AFB	US Air Force Base (on sovereign territory)
AFCD	Advanced Fighter Capability Demonstrator (ex-Strike Eagle)
AGM-	US designation for air-to-ground missile
AIM-	US designation for air intercept missile
AIS	Avionics intermediate shop
AMRAAM	Advanced medium-range air-to-air missile
analogue	Electronic system in which quantities are represented by electrical signals of variable characteristics, i.e. by electrical analogues
anhedral	Angle of wing, canard or tailplane below horizontal
ARI	Aileron-rudder interconnect
ASAT	Anti-satellite (missile system)
aspect ratio	Wing slenderness in plan form, numerically span²/area
ASRAAM	Advanced short range air-to-air missile
ASW	Anti-submarine warfare
augmentation	afterburning (reheat)
AWACS	Airborne warning and control system (Boeing E-3A Sentry)
BIT(E)	Built-in test (equipment)
BVR	Beyond visual range
bypass ratio	Ratio of total airflow through a turbofan engine to that passing through the core section
camber	Curvature of the centreline of a wing aerofoil
CAS	Control augmentation system
CDIP	Continuously displayed impact point
CEP	Circular error probability
CFS	Concept formulation study
chord	Imaginary line joining the leading and trailing edges of a wing or aerofoil section
clutter	Spurious returns on a radar scope
CRT	Cathode ray tube
CW	Continuous-wave radiation
DCP	Development concept paper
DFE	Derivative fighter engine (now the GE F110)
digital	Electronic system in which quantities are as on/off signals coded to represent numbers
dihedral	Angle of wing, canard or tailplane above the horizontal
Doppler	Radar which measures changes in frequency between reflections in the ground ahead of and behind the aircraft, thus giving accurate measure of speed over the ground; Doppler effect is also used to pick out moving targets.
DTD	Damage tolerant design
ECCM	Electronic counter-countermeasures
ECM	Electronic countermeasures
ECP	Engineering change proposals
EO	Electro-optical
EW	Electronic warfare
FAST packs	Fuel and sensor tactical packs (also known as conformal tanks)
FIS	Fighter interception squadron
FLAMR	Forward looking multi-mode radar
FLIR	Forward looking infra-red
FSD	Full scale development
FWW	Fighter weapons wing
FY	Fiscal year (as in US budgets)
g	Acceleration due to standard gravity, unit of linear acceleration
GE	General Electric Company (USA)
GHz	GigaHertz (Hertz × 1,000,000,000)
GPS	Global positioning system
HF	High frequency
HUD	Head-up display
Hz	Hertz, cycles per second
IDFAF	Israeli Defence Force – Air Force
ILS	Instrument landing system
INS	Inertial navigation system
IR	Infra-red, heat radiation
JASDF	Japanese Air Self-Defense Force
JTIDS	Joint tactical information distribution system
KHz	Kilo Hertz (Hertz × 1000)
LANTIRN	Low-altitude navigation and targeting infra-red for night
LLTV	Low-light television
LRU	Line replaceable unit
Mach	Unit equal to the speed of sound
Maple Flag	Series of tactical air exercises carried out over Canada in as realistic manner as possible, including EW
MCAIR	McDonnell Aircraft Company (part of the McDonnell Douglas Corporation)
MMH/FH	Maintenance man hours per flight hour
MSIP	Multi-staged improvement programme
MTBF	Mean time between failure
MW	Megawatt
NDT	Non-destructive testing
PACAF	US Pacific Air Force
Pave Tack	FLIR sensor/laser designator system carried on F-111F and F-4 Phantom, and tested on F-15; produced by Ford Aerospace
PRF	Pulse repetition frequency
PSP	Programmable signal processor
P&W	Pratt & Whitney
Red Flag	Series of tactical air exercises carried out in Nevada in as realistic manner as possible, including EW
RFP	Requests for proposals
RoKAF	Republic of Korea Air Force
RSAF	Royal Saudi Air Force
RWR	Radar warning receiver
R&D	Research and development
SAM	Surface-to-air missile
SAR	Synthetic aperture radar
Seek Talk	Codename for a secure communications system under development
semi-active	Homing on radiation reflected from a target illuminated by radar carried in fighter or other vehicle (but not the missile itself)
slick	Streamlined (Aero-1A shape) bomb
SPR	Sortie production rate
SRAM	Short range attack missile
TAC	USAF Tactical Air Command
TACAN	Tactical air navigation
TEWS	Tactical electronic warfare system
TFS	Tactical fighter squadron
TFTS	Tactical fighter training squadron
TFTW	Tactical fighter training wing
TFW	Tactical fighter wing
UHF	Ultra-high frequency
USAF	United States Air Force
VFAX	Project designation for a carrier-based fighter/attack aircraft
VFX	Project designation for a carrier-based fighter aircraft
VHF	Very-high frequency
Wild Weasel	Codename for defence-suppression aircraft (as in F-4G Wild Weasel)
wing loading	Aircraft weight divided by wing area
'Zulu' alert	Aircraft maintained on constant readiness to scramble against unidentified air targets 365 days a year

Specifications

	F-15A	F-15C	F-15E
Length	63ft 9in/19.43m	63ft 9in/19.43m	63ft 9in/19.43m
Wingspan	42ft 9¾in/13.05m	42ft 9¾in/13.05m	42ft 9¾in/13.05m
Height	18ft 5½in/5.63m	18ft 5½in/5.63m	18ft 5½in/5.63m
Weights			
Empty	28,000lb/12,700kg	28,000lb/12,700kg	31,700lb/14,379kg
Takeoff (air-to-air)	41,500lb/18,824kg	44,500lb/20,185kg	44,823lb/20,331kg
Maximum takeoff	56,000lb/25,401kg	68,000lb/30,844kg	81,000lb/36,741kg
Wing area	608sq ft/56.5sq m	608sq ft/56.5sq m	608sq ft/56.5sq m
Load factor	7.2g/−3g	7.2g/−3g	±9g/−3g
Combat thrust: weight ratio	1.4:1	1.3:1	
Maximum speed	>Mach2.5	>Mach2.5	>Mach2.5
Service ceiling	65,000ft/19,813m	65,000ft/19,813m	65,000ft/19,813m
Range			
Ferry (with external tanks)	>2,500nm/4,630km	>2,500nm/4,630km	
Ferry (with FAST Packs)	—	>3,000nm/5,556nm	
Internal fuel	11,635lb/5,278kg	13,455lb/6,103kg	13,123lb/5,952kg
Number of hardpoints	5	5	
Maximum ordnance load	16,000lb/7,257kg	16,000lb/7,257kg	24,500lb/11,113kg

Picture credits

Page 4: both: US Department of Defense. **5:** all: McDonnell Douglas. **6:** all: McDonnell Douglas. **7:** both: McDonnell Douglas. **8:** both: USAF. **9:** both: McDonnell Douglas. **9:** McDonnell Douglas. **10:** both: McDonnell Douglas. **12:** top: McDonnell Douglas; bottom: US Department of Defense. **13:** McDonnell Douglas. **14:** top: US Department of Defense; bottom: McDonnell Douglas. **15:** both: McDonnell Douglas. **16:** all: McDonnell Douglas. **17:** top left and right: USAF; bottom: McDonnell Douglas. **18:** Lindsay Peacock. **19:** top: US Department of Defense; below: Lindsay Peacock. **21:** top left: McDonnell Douglas; top right: Lindsay Peacock. **22:** top: USAF; bottom: McDonnell Douglas. **23:** all: Pratt & Whitney. **24:** top: USAF; bottom: McDonnell Douglas. **25:** top: US Department of Defense; bottom: USAF. **26:** top: US Department of Defense; bottom: General Electric. **27:** all: Pratt & Whitney. **28:** top: McDonnell Douglas. Bottom: Lindsay Peacock. **29:** McDonnell Douglas. **30:** McDonnell Douglas. **31:** top and bottom: McDonnell Douglas; centre: USAF. **32:** McDonnell Douglas. **33:** Hughes Aircraft. **34:** top: US Department of Defense; centre: McDonnell Douglas. **36:** top and centre: McDonnell Douglas; bottom: USAF. **37:** top: General Electric; bottom: McDonnell Douglas. **38:** all: McDonnell Douglas. **39:** top and upper centre: US Department of Defense; lower centre, bottom left and right: Hughes Aircraft. **41:** both: McDonnell Douglas. **42:** top: USAF; bottom: McDonnell Douglas. **43:** top and centre right: McDonnell Douglas; bottom left: US Department of Defense; bottom right: USAF. **44:** both: USAF. **45:** top: US Department of Defense; bottom: McDonnell Douglas. **46:** all: US Department of Defense. **47:** all: US Department of Defense. **48:** US Navy. **49:** both: US Department of Defense. **50:** McDonnell Douglas. **51:** McDonnell Douglas. **52:** top: USAF; bottom: McDonnell Douglas. **53:** top: US Department of Defense; bottom: McDonnell Douglas. **54:** McDonnell Douglas. **55:** top: US Department of Defense; bottom: McDonnell Douglas. **56:** top: McDonnell Douglas. **56-7:** US Department of Defense. **57:** top: McDonnell Douglas; centre: USAF. **58:** top and bottom: McDonnell Douglas; centre: Matra. **59:** both: McDonnell Douglas. **60:** top: McDonnell Douglas; bottom: Pratt & Whitney. **61:** both: McDonnell Douglas. **62-63:** all: McDonnell Douglas.